# Praise for *Science of the Magical*

"Kaplan takes us on a journey spiced with the wonders of myth, history, and art, leavened with impeccable research, endlessly fascinating. And the result is both a compelling read and a deeply thoughtful exploration of the world around us and the ways we seek to understand it."

—Deborah Blum, author of *The Poisoner's Handbook*

"Erudite, witty, and highly original . . . Kaplan will not only enlighten and charm you but also change the way you think about what is science and what is magic."

—Daniel Lieberman, professor of human evolutionary biology, Harvard University, and author of *The Story of the Human Body*

"Nimbly explores topics as diverse as berserkers, hepatomancy, Methuselah mice, stage magic, superheroes, and sunstones, taking the reader on wide-ranging journeys from Iceland, Turkey, and Yellowstone National Park to backstage in Las Vegas in his search for the science behind magic and myths."

—Stephen A. Mitchell, professor of Scandinavian and folklore, Harvard University, and author of *Witchcraft and Magic in the Nordic Middle Ages*

"A wonderful exploration of the possible. Kaplan takes us on a journey into the myths and magic of days long gone and then looks to make scientific sense of these ancient ideas and practices."

—Jeffrey Shaman, associate professor of environmental health sciences, Columbia University

"A relevant, fascinating exploration of just how weird the world can be. I tell people about things I read in it at least once a week. In a world of bad news, it's refreshing to pick up a book that is filled with wonder, mystery, and joy, especially when all of that comes steeped in scientific evidence."

—Maggie Koerth-Baker, journalist and author of *Before the Lights Go Out: Conquering the Energy Crisis Before It Conquers Us*

"This book reminded me of all the fascination and excitement I felt about the universe and how important it was for me to study life and our place in the world. In short, it made me remember why I became a scientist."

—Elizabeth Repasky, professor of oncology, Roswell Park Cancer Institute, New York

## ALSO BY MATT KAPLAN

*The Science of Monsters: The Origins of the Creatures We Love to Fear*
(previously published as *Medusa's Gaze and Vampire's Bite*)

# SCIENCE

## of the

# MAGICAL

---

from the **HOLY GRAIL**
to **LOVE POTIONS**
to **SUPERPOWERS**

---

# MATT KAPLAN

Scribner

New York   London   Toronto   Sydney   New Delhi

Scribner
An Imprint of Simon & Schuster, Inc.
1230 Avenue of the Americas
New York, NY 10020

The jacket art for this edition of *Science of the Magical* is taken from a lithograph of a patent medicine* label stored in the Library of Congress: "Sybilline Leaves by Florence & Fanny, Philadelphia / T. Sinclair's Lith., Phila., c. 1852."

First Scribner hardcover edition October 2015

SCRIBNER and design are registered trademarks of The Gale Group, Inc., used under license by Simon & Schuster, Inc., the publisher of this work.

For information about special discounts for bulk purchases, please contact Simon & Schuster Special Sales at 1-866-506-1949 or business@simonandschuster.com.

The Simon & Schuster Speakers Bureau can bring authors to your live event. For more information or to book an event, contact the Simon & Schuster Speakers Bureau at 1-866-248-3049 or visit our website at www.simonspeakers.com.

Interior design by Jill Putorti

Manufactured in the United States of America

10  9  8  7  6  5  4  3  2  1

Library of Congress Cataloging-in-Publication Data is available.

ISBN 978-1-4767-7710-8
ISBN 978-1-4767-7713-9 (ebook)

---

* "Patent medicine" is just a polite way of referring to the various potions and drugs sold by ethically questionable peddlers through the ages. Sometimes these products actually produced the miraculous effects that the salesmen claimed they did, and sometimes they did not.

*In loving memory of Patricia Morse*
*for helping me to reach* The Hobbit *when*
*I was too small to reach it by myself.*

# CONTENTS

# INTRODUCTION

*And some things that should not have been forgotten, were lost.*
*History became legend and legend became myth.*

—GALADRIEL, *THE FELLOWSHIP OF THE RING*

The Egyptian war chariots were fast approaching. With no weapons, few rations, and the scorching desert sun beating down upon them, the situation was looking dire for the fleeing Hebrews. Then things went from bad to worse as they found themselves at the edge of the Red Sea. Capture seemed inevitable. The chariots drew closer; all seemed lost; then it happened. Moses lifted up his staff and called upon God to aid the people in their time of need—and help them God did. The waters parted, the Hebrews ran to safety, and the Egyptians were drowned by the crashing waves as they tried to follow.[1]

Told over and over through the generations and depicted in countless works of art, the parting of the Red Sea in Exodus is one of the most gripping supernatural moments in Western mythology. It is also an event that, like the ten plagues, archaeologists, historians, and Bible scholars have relentlessly dissected as they have sought to determine if fragments of fact are nestled among the fiction. Might such a story contain descriptions of natural events like earthquakes, floods, or storms that our ancestors witnessed but could not understand? Yes, but we need not only look to our most ancient myths to find bits of recorded history.

Several years after accidentally being exposed to a high dose of radiation while working at a nuclear power plant, engineer Norton McCoy fathers a son named Hank. As the boy grows, he develops inhumanly long limbs, incredible strength, and apelike hands and feet. Due to his unusual body, he comes to excel at numerous sports at school but quickly meets with discrimination from other students on account of his being different. Hank soon discovers that he carries mutant DNA on account of his father's radiation exposure, and while he is not alone, he is a minority in a world filled with people who both hate and fear him.[2] He soon changes his name to a more appropriate alias, Beast, and ultimately joins the X-Men to fight for mutant rights.

There is little doubt about what realities the *X-Men* comics were recording when they were first written in 1963. The Civil Rights Act was only months away from being signed into law by President Lyndon Johnson. The issue of discrimination was on everyone's mind, and the persecution of mutants as a metaphor for this issue was ideal. Yet not just racial tensions were captured by this modern mythology. Science was recorded too.

Mutants were initially called the "children of the atom" for a good reason. The Cuban Missile Crisis, which is depicted in the movie *X-Men: First Class*, took place just a year before the comics were written. Anxiety about the effects of radiation on human biology was at an all-time high and was preserved in the *X-Men* comic series. The Hulk is much the same, appearing in 1962 and described as the result of gamma radiation dramatically altering Bruce Banner's body. Spider-Man, also created in 1962, similarly gains his powers after being bitten by an irradiated spider.[3]

It may seem jarring to throw comic-book superpowers into the same bin as the magical events of Exodus. From an early age most of us are taught to see magical acts performed by gods as somehow different from magical acts performed by mortals. But are they really all that different? Whether we call it divine intervention, a miracle, the supernatural, sorcery, or mutation, all of these things share a core similarity. They present the impossible as real while simultaneously recording information about what people were experiencing at the time these stories were created.

## FANTASY FOSSILIZED

We have an insatiable appetite for comprehending the world around us. When we see things that we do not understand, our brains immediately get to work trying to make sense of them. These can be minor matters, such as noticing that you sleep less well at the time of the full moon than you do when it is a crescent. They can also be monumental, such as a nonbreathing and pulseless patient suddenly coming back to life after five minutes of effectively lying dead on the surgery table.

Fascinated and baffled, we find ourselves wondering. Is it really the full moon that is making me sleep so poorly,* or is it something that I'm doing at that time that is causing me so much trouble? Was there enough glucose and oxygen stored in that patient's capillaries to keep the brain from suffering permanent damage during those five minutes? Was the person in some sort of deep hibernation and not actually dead? When none of these explanations seem acceptable, we turn to the supernatural. The moon casts a spell upon us. An angel stepped in and guided the departing spirit back to the living world.

One of the most dramatic, and tragic, examples of this phenomenon comes from the world's recent struggle with Ebola. Patient Zero, the first person during the 2013–2014 epidemic to catch the disease, was a little boy named Emile living in the rural village of Meliandou, Guinea. He quickly developed a high fever with vomiting and bloody stools and

---

* Just in case you were wondering, it is the full moon and not that shot of espresso you had before dinner. Well, the shot wouldn't help, but the moon does seem to mess with our sleep patterns. More on that later, though.

died. A few days later, his sister caught the disease and passed away as well. Shortly thereafter, Emile's pregnant mother fell ill and started bleeding heavily. Late in the night she suffered a miscarriage and died. Three women from the village came to clean up the mother's blood, and they too perished. Only Emile's father survived. Regional medical clinics were baffled. Locals were terrified. In the wake of all the fear and uncertainty, the people of the village—including the ill—came together to perform rituals that they believed would protect them against the black magic of the curse that had struck their community. These magical rituals were a terrible mistake. Multiple new cases followed the ceremony, and the disease spread like wildfire from there.[4]

The situation for our ancestors was much the same. While many of us today look at the fiercest of thunderstorms and explain the chaos as the result of pressure systems and temperature changes, our ancestors were not equipped with such information. They looked at the lightning bolts, deafening thunder, and devastating hail and came to the only conclusion that they reasonably could: Thor, Baal, or Zeus was angry. Similarly, when our ancestors found the bones of fish and the shells of clams stuck in the rocks of mountains thousands of feet above sea level, it would have been reasonable for them to speculate that there had once been a great flood. Yet it would be wrong to always portray people who lived long ago as the clueless ones. Things have sometimes worked in reverse, with our ancestors understanding the world in a remarkable way that has been lost, or very nearly lost, to the ravages of time.

Legend tells of the Vikings possessing an artifact called the sunstone, which allowed them to successfully navigate the Atlantic Ocean centuries before the invention of the magnetic compass. For decades, historians dismissed the sunstone as mere fantasy, but evidence is now emerging from the fields of physics, mineralogy, and archaeology that this object actually existed.[5] Similarly, stories in *The Odyssey* portray the Greeks as aware of plants with powerful medicinal properties. They considered them to be magical herbs and often connected them to fantastic stories of the gods. For centuries such tales were disregarded as nothing more than fictions, but recently a number of scientists and historians have started to think that the Greeks were onto something.[6]

As these examples and the many others in this book suggest, magic can function a bit like a fossil. Just as we can look at the bones of animals that lived long ago and use the evidence to deduce what the past was like, we can look at the magic of our ancestors to hypothesize about what they may have dreamed of and what they might have seen in the world around them. In a sense, Galadriel's words were accurate: history can become legend and legend can become myth.[7] Nevertheless, it would be wrong to suggest that all magic arises from our struggle to comprehend the complexities of the surrounding world. Sometimes we just imagine amazing things and then try to make them real.

## PRACTICAL MAGIC

Guided by beliefs and dreams, we have a natural tendency to try to make magical things happen. I didn't try parting any bodies of water as a child, but on a fair few boring afternoons in elementary school, as a *Star Wars* junkie, I tried to use the Force to move a pencil on my desk just a few inches closer. . . . It never worked. On a more somber note, I vividly remember sitting at my grandmother's grave shortly after her death and desperately wishing I could speak with her one last time. Of course, I am hardly alone.

Every year at Halloween hundreds of people gather at Harry Houdini's former house expecting him to one day find a way to escape death. Hundreds of thousands carry around "lucky" rabbit's feet,* countless

---

* Incidentally, the ever-vigilant gaze of the rabbit was thought to provide protection against the "evil eye." However, rabbit eyeballs don't do too well after a few days on the end of a key chain. In contrast, fluffy feet hold up for years and have the added benefit of retaining cheap pink dyes rather nicely.

millions pray for God to intervene in their lives, and newspapers around the globe still print astrology sections that many take seriously. None of these behaviors are new.

Long ago, people put requests for the gods on lead tablets and tossed these messages into sacred pools, put wax dolls resembling real people into sexually explicit positions to create feelings of lust in targeted individuals, or, my personal favorite, put nails through the hearts of chickens stuffed with hairs from the heads of enemies and left these abominations on doorsteps to cast curses.[8] Sounds about as divorced from science as you could possibly imagine, and in many cases this stuff was utter nonsense . . . but not always.

Healing rituals in ancient temples involved some techniques that we now know yield health benefits; drinking out of holy grails crafted from specific rocks found in the caves where saints lived may truly have granted a longer life under certain circumstances; and the search for the stone of ever-lasting life, known as the philosopher's stone, led to major findings in what would eventually become chemistry.[9]

To this day, illusions drawn by artists such as Maurits Cornelis (M. C.) Escher and stage performances by modern magicians are helping to guide neuroscientists as they figure out how the human brain makes sense of the complicated world that we live in.[10] So while much magic tiptoes its way into the world of science, a lot of science pushes—or, dare I say it, bulldozes—its way into the world of magic.

Seeing in the dark was once an ability that only cats and sorcerers could wield; now anyone with the right set of goggles can manage it. Reading minds was once the territory of oracles and Charles Xavier, but technology is on the verge of granting us the ability to detect what completely paralyzed patients who have lost motor control of their mouths are thinking. In days long gone, the idea of flying on a carpet or broom was strictly the stuff of fantasy. Now we have individuals, such as Yves Rossy of Switzerland, who dart through the skies via jet-propelled devices similar to the flight system used by Tony Stark in *Iron Man*.

Such developments are not limited merely to the world of engineering. While love potions were once only found in Shakespearean plays and the workshops of witches, we now have a veritable cornucopia of pharmaceuticals that can do everything from messing with sexual arousal to making two people more likely to become friends during an initial encounter. On a more frightening level, while it was once the remit of the Viking god Odin to possess the minds of his followers on the field of battle and transform them into the raging berserker warriors of legend, we are now starting to understand which compounds these men were consuming and have the potential to use them to create biochemically altered soldiers who feel neither pain nor fear. Many ethical issues arise here for sure, but all of these magical transitions into reality raise a much larger question about the nature of magic itself.

Does knowing how something magical works make it into something other than magic? I had no idea when I set out to write this book. I had a long love of all things magical found in our mythology, but as a paleontologist by training and a science journalist by profession, I also had a passion for understanding the realities underlying complex systems. After spending two years poring over the myths, beliefs, and rituals of our ancestors, I was still unsure. The question itself proved cloaked in illusion. If I considered just one or two isolated examples, it seemed deceptively simple to answer. However, the more I pondered the matter, the harder it became to grapple with.

Long ago, when faced with perplexing questions, our ancestors trekked for hundreds of miles to gain wisdom from the great oracle at Delphi. When the FASTEN SEAT BELTS sign illuminated on my flight from London to Las Vegas, I realized that I was on a similar pilgrimage. To truly understand how knowing shaped magic, I needed the help of a magician.

Onstage, Teller may be the silent half of the magic and comedy duo Penn & Teller, but he is hardly silent behind the scenes. He has a sharp philosophical mind, and while he speaks quietly compared to his towering companion, his every word is valuable.

As I asked him about the relationship between knowing and magic, he quoted his magician friend Mike Close as saying, "Magic is the gift of a stone in your shoe. You leave with something that you can't quite figure out and just can't stop thinking about." That seemed to fit perfectly with what I knew of modern stage magicians. Keeping audiences guessing was pivotal to their profession. However, just moments later, Teller added, "To any enlightened dweller of our century, knowing increases wonder. I don't just think this, I know it. If you believe, you oversimplify in the way a child might initially believe there is no complex evolution of life. That simplification in no way increases your sense of wonder. If, however, you *know* how life came to be on our planet, the wonder is immense!"

His answers were a riddle worthy of the oracle herself. If magic depended upon a viewer's not knowing how the magic happened, and if knowing was directly tied to a sense of wonder, was Teller saying that being duped by a trick was not as wondrous as knowing how the trick was done? That couldn't be right. Could it?

What the answer is, or whether there even is one, will, I hope, become clear as we embark on this exploration of the ways in which magic through the ages has been wielded, crept into our stories, guided our research, and in many cases come to be much more than myth.

# HEALING

Prayers, Sacred Pools, Regeneration, and Holy Eyeliner

*The healing power of the Grail is the only thing that can save your father now.*
*It's time to ask yourself what you believe.*

—WALTER DONOVAN, *INDIANA JONES AND THE LAST CRUSADE*

Entering our bodies, viruses and bacteria multiply rapidly. Our immune systems often take action fast enough to thwart their spread, but sometimes the pathogens get the upper hand and we fall ill. While symptoms vary, they are universally unpleasant. Gastrointestinal infections leave us heaving in the bathroom. Upper-respiratory bugs cause sore throats, blocked-up noses, plugged ears, and hacking coughs. Influenza leaves us sapped of energy, feverish, and racked with pain. Yet we have it rather easy.

Today we have effective tools at our disposal ranging from Tylenol and aspirin to antibiotics and antivirals that allow us to fight back against pathogens. True, bacteria that are resistant to our antibiotics are evolving, and our antivirals are still limited in their effectiveness; but we are far better off now than we were sixty years ago.

Shortly after getting back from the South Pacific at the end of World War II, my grandfather contracted polio. Talk about rotten luck. He'd survived two years of bombs and bullets only to be taken down by a virus just as the vaccine started being distributed. Fortunately, his doctors were able to keep him alive with an iron lung as the paralysis took temporary hold of his respiratory system, but what a horror to know that the virus was spreading with no way to fight it. A couple thousand years ago our ancestors had it even worse.

Centuries ago, humanity not only had few treatments for diseases but also had little idea why people became ill in the first place. Every bout of sickness was scary. While the very young and very old were the most at risk, being in the prime of life did not come with the sense of invulnerability that it does today. With everything from smallpox to tuberculosis running rampant, anyone could die at any time.

The sheer randomness of it all drove frightened people to seek explanations, and this search led to an exploration of the supernatural. We know that ancient populations made sacrifices and prayed to the gods when they sought healing. As unscientific as these actions might seem, they did have a sort of logic.

Seizures baffled our ancestors for centuries. If a patient was seen to be imitating a goat during a seizure by grinding his or her teeth or having convulsions on the right side, it was deemed that Zeus's mother, Rhea, was responsible for the affliction, and attention was given to the goddess to help alleviate her wrath.* If a patient passed feces that were thin, like those made by a bird, prayers were made to Apollo, the god of the sun and the skies, where birds fly high.[1] A foaming mouth or frantic kicking of the feet indicated that Ares, the god of war, was behind the seizure. Delirium? No problem. It had to be a spell cast by Hecate, the goddess of sorcery.[2] The list went on and on.

---

* Zeus's father, Cronus, had a bad habit of swallowing whole the children that Rhea birthed him. After Hades and Poseidon went down the hatch, Rhea was fed up with having her kids eaten, so she handed Cronus a stone wrapped in a baby blanket and told him it was his son Zeus. Being none too bright, Cronus swallowed the stone and never noticed that Rhea had secreted Zeus off to a cave on the island of Crete, where he was raised on the milk of goats. It is admittedly a stretch, but this is where classicists tell us the Rhea-goat connection comes from.

It is easy to sit back and smile in amusement at the behaviors of our ancestors, but ask yourself, were the Greeks really behaving all that differently from many people today? When I was doing rounds in the emergency room as a trainee emergency medic back in 1998, I heard more than a fair few people say, "I'll be praying for you," as their loved ones were wheeled off to the operating room. According to the New Testament, Christ brought sight to the blind, gave movement to those suffering from paralysis, and even cured patients suffering from the disfiguring bacterial disease leprosy.[3]

Jesus was hardly alone. Dozens of deities in human form and prophets from the Christian, Jewish, Islamic, Hindu, and Buddhist faiths are said to have performed similar deeds. Could there actually be something behind all of this mythology? Can belief or faith have any measurable effects on human health?

## THINKING YOURSELF WELL

The link between mind and health is an academic minefield that many researchers, fearing ostracism, have dared not tread. Yet much is worth studying in this murky territory. Best known is the placebo effect, in which the mere belief by a patient that he or she is being treated leads to improvements in health, even if the treatment is nothing more than an injection of saline solution or a tablet of sugar. In the past decade, however, researchers have also studied the relationship between positive emotions and the body.

One study, conducted in 2006, showed that experiencing warm and upbeat emotions resulted in patients experiencing fewer colds.[4] Another study, conducted in 2007, discovered that such positive feelings were connected to reduced systemwide inflammation.[5] A 2012 paper even provided evidence that frequent happiness correlated with reduced cardiovascular disease.[6] The message seemed clear. But were these studies really finding that being happy led to improved health? It seemed prudent to speak with an expert. So I soon found myself sitting in an office at the University of North Carolina, Chapel Hill.

"We were curious if these were just correlations," explained psychologist Barbara Fredrickson, an expert on positive emotions. "Nobody has been sure whether people who were naturally inclined to be happy tended to experience fewer colds, less inflammation, and less heart disease, or if the actual process of being made happy had this effect." Keen to explore this, she ran an experiment and found that individuals randomly assigned to generate positive emotions reported experiencing fewer headaches and less chest pain, congestion, and overall weakness.

This was interesting, but it didn't explain why being happy could affect health. "We needed an experiment that accurately measured the ways in which positive emotions were actually altering how the human body functioned," explained Dr. Fredrickson. With this in mind, she designed a new experiment that concentrated attention on the vagus nerve.

The vagus nerve starts in the brain and runs, via multiple branches, to numerous vital organs.[7] One of these organs is the heart. Among the vagus nerve's many jobs is to send signals that tell the heart to slow down when no danger is in the surrounding area. Generally, the vagus nerve is considered to be healthy if the heart rate subtly increases while one breathes in and subtly decreases while one breathes out.[8] The difference between these two is called the vagal tone. High vagal tones are associated with overall good health and few cardiovascular complications, while low values are linked to inflamed tissues and heart attacks.[9]

Interestingly, those with a high vagal tone were known to be better at keeping their negative feelings from getting overblown than those with a low vagal tone.[10] High vagal tone also seemed to be connected to the presence of more positive emotions overall. Armed with all of this information, the researchers measured the vagal tones of sixty-five participants at the beginning of the nine-week experiment and then divided them into two groups at random. Half were taught a meditation technique known to create feelings of goodwill both toward others and oneself and asked to meditate daily.[11] All participants went to a website every evening to rate, on a five-point scale, whether, and how strongly, during the day they felt nine positive emotions, such as hope, joy, and love, and eleven negative ones, including anger, boredom, and disgust.[12] At the end of the experiment, all had their vagal tone measured once more.

While the team was unsurprised to see that positive emotions increased among those participants who meditated, they *were* surprised that the participants' vagal tone went up too.[13] In contrast, emotions and vagal tone remained unchanged in the group that did not meditate.[14] These findings, published in the journal *Psychological Science*, hinted that something about the feelings of hope, love, and joy was having a physical impact on health.

The findings provided more questions than answers. Beyond vagal tone, what was happening to the study participants at the biochemical and molecular levels? Did all positive emotions have the same biological effects, or were some more powerful than others? Eager to explore these questions more closely, Dr. Fredrickson designed an experiment in collaboration with a team of genomicists led by Steven Cole at the University of California, Los Angeles.

The experiment took aim at the two types of happiness: *eudaimonic* happiness, which stems from doing virtuous things, such as feeding the hungry, helping the elderly, donating to charity, and just generally being a good person, and *hedonic* happiness, which is the happiness that one gets from spending a weekend in Las Vegas. Dr. Fredrickson and Dr. Cole wondered if these forms of happiness functioned differently at the biological level.

The researchers asked eighty-four healthy volunteers a series of questions like "How often did you feel satisfied in the past week?" and "How often did you feel that your life had a sense of direction or meaning to it?"* All participants provided the researchers with twenty-milliliter blood samples, which were analyzed for gene expression.[15] The results were astonishing.

Dr. Fredrickson and Dr. Cole reported in 2013 in the *Proceedings of the National Academy of Sciences* that, biologically, the two forms of happiness could not be more different.[16] Participants who scored high on eudaimonic happiness and low on hedonic happiness showed 10 percent greater expression of the genes associated with the production of interferons, proteins that support communication during immune system responses, and a 30 percent greater expression of the genes associated with the production of proteins that defend against pathogens.[17] In contrast, participants who scored high on hedonic happiness and low on eudaimonic happiness showed 20 percent greater expression of inflammation-causing genes in their blood samples and 20 percent less expression of genes associated with the production of proteins that defend against pathogens.[18] To put it more simply, the happiness that arises from being a noble soul brings improved immune system response and better protection against pathogens, while happiness associated with selfish pleasures brings less protection against pathogens and greater inflammation.**

In fairness, the experiment looked at the gene activity of people who were rather extreme, i.e., very eudaimonic and not very hedonic, or the other way around. For most people, who regularly experience both forms of happiness, it is likely the two effects cancel one another out. It

---

* Asking participants to report how they feel is always a challenge for psychologists. For example, problems can arise if a participant in a study badly *wants* their life to have meaning or believes they should feel satisfied after an activity. Under these circumstances they can report feeling a certain way when they actually do not, and that can skew results. Researchers try to work with lots of people and study the responses for reliability, as was done with this study, to help ameliorate this problem, but keep in mind that the method isn't perfect.

** While chronic inflammation causes many ills, short-term inflammation plays an important role as an immune system accelerator and helps to rapidly heal wounds. Thus, under certain circumstances, it may still be possible to justify a weekend in Vegas as important for improving one's health.

is also theoretically possible that causation runs in the opposite direction, and that people with particular patterns of gene expression are healthier, take a longer-term view of life, and are thus living more noble and charitable lives. Dr. Fredrickson's research with vagal tone and gene expression in association with happiness was replicated twice in 2015, but more work is still needed to tease such complex questions apart.[19] Nevertheless, it looks increasingly likely that the link between the mind and physical health is more important than we ever realized.

After conducting his many miraculous healings, Jesus Christ often said, "Your faith has healed you."[20] Whether these healing events actually happened is not for me to say. I wasn't there and am neither a historian nor a theologian. However, I wondered whether there could be a kernel of truth here. Fascinated and unsure, I gave Dr. Fredrickson a call.

"Might there be something real to these biblical stories? Could the act of praying to a god or engaging in some sort of sacred act have actual physiological effects?" I asked.

Dr. Fredrickson sighed quietly on the other end of the line. I could almost hear her mind pondering. She knew my science writing in the *Economist* well. We'd collaborated on a couple of stories over the years, and she knew I wasn't going to twist her words into something outlandish. I was bringing up a delicate topic and knew she was choosing her words carefully.

"Taking your last bit of life, directing it toward something socially important, making a pilgrimage, pinning your fate to this fabric of gods and people that was so much larger than yourself . . ." She paused and took a breath. "There wouldn't have been any instant healing, like what you read about in the myths, but you would expect some real immunological effects from all this deeply meaningful stuff."

She explained that this relationship between emotion and health is actually encouraging. While our ancestors believed the gods held their lives in their hands, the reverse is the case. "There's this idea that emotions just rain down on us, but that's not really true. While some people are genetically predisposed to be negative, they still have the power to make themselves happy and gain health benefits; it just takes them more effort to get there."

Did our ancestors, while they made their sacrifices and prayed with all their hearts, have any sense that the positive emotions associated with noble, community-directed behaviors played a part in boosting their immune systems? We can't know that for certain. But the fact that gods, temples, pilgrimages, faith, and laying-on-of-hands rituals have been tightly connected to healing for centuries leads me to suspect that our ancestors were aware of far more than we realize.

## PROTECTIVE PIGMENTS

Magical healing was not strictly limited to gaining the attention of the gods through prayer. It sometimes involved other practices, such as applying eye shadow in just the right way.

In Egypt, green and black makeup was extensively applied around the eyes by followers of Horus, the god of the sky and kings. We know from ancient records that this makeup was often a sign of devotion. Horus had the head of a falcon, and most falcons have dark stripes near and around their eyes.[21] The faithful of Horus were emulating their god's appearance by painting their faces in a similar way.[22] Moreover, this activity was not limited to just the priests of the god. A number of archaeological finds depict commoners who also had their eyes painted with dark green and black.[23] None of this was surprising when it was discovered during the past century. People have a long history of using costumes and makeup that help them to become better connected with their deities. What was a surprise to archaeologists were the descriptions of the magical effects the Egyptians believed application of this makeup granted them.[24]

Horus was mortal enemies with Set, the god of storms and war, and according to mythology, in several battles Set wounded Horus in the eye.[25] This partial blinding of the sky god put the world at risk of falling into eternal darkness. But other gods always came and poured magical substances into Horus's eye to help repair it.[26] The eye of Horus became a symbol of healing, and so too did the makeup that his followers wore.[27] The only problem with this belief was the chemistry behind it. When jars of the makeup found in Egyptian tombs were analyzed, lead turned up.

Lead is toxic, and the eye shadow was made up of lead chlorides such as laurionite, $Pb(OH)Cl$, and phosgenite, $Pb_2Cl_2CO_3$. For decades, modern scholars assumed that the followers of Horus were unknowingly poisoning themselves.[28] Then, in 2010, a team led by chemists Christian Amatore at the National Center for Scientific Research in France and Philippe Walter at the Louvre revealed that there was more to the makeup than anyone could have imagined.

Striking similarities exist between the lead ion $Pb^{2+}$, which would readily be shed by laurionite and phosgenite, and the calcium ion $Ca^{2+}$, which plays a key role in kicking two major parts of the immune system into gear.[29] $Pb^{2+}$ and $Ca^{2+}$ have identical charges, identical coordination numbers with water molecules, and similar atomic radii.[30] Moreover, because most lead compounds found in nature do not dissolve and have little ability to release $Pb^{2+}$ ions into the body, the researchers hypothesized that cells along the thin tissues of the eye might mistake the ions shed by laurionite and phosgenite for $Ca^{2+}$ and respond accordingly.[31] Keen to find out, they ran an experiment.

The researchers exposed skin cells called keratinocytes, which they cultured in the lab, to solutions containing tiny levels of $Pb^{2+}$,* the same ions that would be found in lead makeup. The cells responded by promptly increasing their production of nitric oxide, a compound that summons phagocytes—immune cells that engulf invading bacteria and destroy them—by a whopping 240 percent.[32] Crucially, Dr. Amatore

---

* For all you biochemists out there who just *have* to know, it was a 0.2 micromolar solution of $Pb^{2+}$ ions.

did not see the low levels of lead change the behavior of the keratinocytes in any other ways.

The finding, which the researchers published in the journal *Analytical Chemistry* in January of 2010, suggested that the Horus makeup was
indeed granting a form of protection.[33] By exposing the keratinocytes
surrounding their eyes to low levels of lead ions daily, Dr. Amatore and
his colleagues argued that the Egyptians were creating a standing army
of phagocytes in that part of their body. Would this have had any real
effect on health that the Egyptians would have noticed? I wasn't so sure,
so I arranged to discuss the matter over a drink with Harvard Medical
School immunologist Denise Faustman.

"I'll tell you, Matt, when you passed this paper along, I initially
thought, no way. I couldn't see how anyone could prove that this makeup
was worn, let alone that it had any immunological effects, but it's convincing. There is no question that a two hundred and forty percent
increase in nitric oxide production is huge. That would summon lots of
cells to fight bacteria," explained Dr. Faustman. Her feedback was reas-
suring, but the critical question that remained on my mind was whether
having an army of immune cells around the eyes would have mattered.

"It would prevent all sorts of eye diseases, especially if these people
were washing their faces in nasty water contaminated with animal waste
from the Nile all the time. A lot of these diseases can make you go blind
if they don't get treated, so preventing them with a topical aseptic like this
would have probably mattered quite a lot. Frankly, nonspecific antibiotics like this are very hard for bugs to adapt to; there may be real value in
studying this stuff more closely for modern use," explained Dr. Faustman.

Dr. Amatore concurs on this point too. "The antibiotics that we use
today cannot kill all bacteria, especially once bacterial populations have
radically expanded. After dividing a lot, bacteria create a lot of genetic
diversity that allows for antibiotic resistance to arise, which makes the
population much harder to kill off. In this respect, the way the makeup

prevents bacteria from settling in the eye in the first place is more effective than many of our present antibiotic strategies."

It all left me wondering, did the Egyptians wear this eye makeup for centuries solely for fashion and religious reasons, or did people notice the protective power of the eye shadow and come to the conclusion that putting it on drew blessings from Horus that kept their eyes healthy?

## SACRED SPRINGS

Drawing the attention of gods was only part of most ancient healing practices. Sufferers also pilgrimaged to holy places. At many of these sites, practitioners undertook what we call incubation therapy, whereby they would fall asleep and be visited in their dreams by the divine being who was thought to be connected to the place. Some believed the supernatural entities sent healing information in dreams that patients could use to help themselves get better, while others believed the entities themselves took direct action to heal worthy individuals. For the devout, sleeping in holy places and being told they were being watched by gods might have triggered some of the positive emotions that do activate the immune system. But some methods used in these sacred places likely had other healing effects as well.

During the height of the Roman Empire people came from across Europe to visit the hot springs at the English city of Bath. Many sought the healing powers of the goddess Minerva, to whom the site was sacred. While Minerva—or Athena, as she was known to the Greeks— is popularly known today as the goddess of wisdom and battle tactics, healing was also once part of her divine portfolio. So sick people flocked to her favored grounds to be granted help.

To this day, millions of people still seek healing from flowing springs at sites such as Lourdes, on the French side of the Pyrenees, where the Virgin Mary is said to have appeared numerous times during the late 1800s. I unwittingly became one of these pilgrims in 2001, during a tough fourteen-day trek through the mountains there. The weather had been foul, and a quarter of our group was suffering from a gastrointestinal disease from improperly treated water. Many were going hungry too, since poor planning had led us to run out of food. To make matters worse, a mix of extreme heat, glacier crossings, and torrential downpours had left a number of us with horrendous blisters.

I was in a sorry state at the time, but I vividly recall two things upon reaching Lourdes. First, a nice lady on the street was selling gelato as we walked into town. Second, locals claimed a spring near our campsite was holy. I had no idea then about the legends associated with the Virgin Mary, but I did offer up my own little prayer as I soaked my destroyed feet. I don't know if I was blessed with any supernatural healing, but it sure felt nice. Our ancestors, however, went much further than just dipping their feet in sacred water.

The healing practices at Bath involved drinking the water and using the hot pools, cold plunges, saunas, and steam chambers. Curious about these activities, I went to Bath with my eager and inquisitive art-historian mother, who was visiting us from the States.

Like many Americans working in London, I'd heard lots about Bath but had never bothered doing something so touristy as to visit. Mom's trip and research for this book proved a good reason to make the journey. With a veritable labyrinth of rooms and passages, the structure at Bath must have been seriously impressive when it functioned as a temple. While the historic facility is no longer open for use, water from the spring at the site can still be imbibed at a few select locations.

"Huh . . . it tastes like water," commented my mother in disappointment. I had to agree. After touring the ruins I was expecting to taste something unique. While many people have argued that the water carries minerals that are good for the health, chemical analysis of it in recent decades has debunked this suggestion. The water, while rich in calcium and sulfate, has nothing particularly medicinal about it.[34]

Eager to interact with the waters as our ancestors did, I booked us the full Roman experience. We had massages that used historical techniques, used a hot room that was filled with hay and smelled like a musty barn, and bathed in pools filled with water from the local springs. I was surprised to discover that the pools were more lukewarm than hot. They were also filled with romantically inclined teens who seemed most interested in plunging their tongues down one another's throats. It was all rather surreal,* and it made me wonder, did our ancestors really believe there was something to all of these activities? The ancient-Greek doctor Hippocrates, who eschewed most of the ideas that the gods caused illnesses and granted health, certainly thought so.

Hippocrates believed that the body was composed of four elements known as humors: black bile, white bile, blood, and phlegm. When the humors were balanced, the body functioned properly; but when one humor grew too great or diminished too much, balance was lost and illness followed.[35] To Hippocrates, getting the humors back into balance was crucial to improving the health of the ill, and a key way of doing this was to manage the environment of the patient. By exposing the ill to various mixes of hot, cold, wet, and dry, he felt that the balance of the humors could be adjusted and health restored. Later doctors, such as Galen, Asclepiades, and Celsus, largely agreed with these ideas, but started to take things further by prescribing specific spa treatments for different diseases.[36] By AD 100, many spa therapies were used to treat a variety of disorders, including paralysis, joint pains, digestive problems, fatigue, skin conditions, and fevers.[37]

Some real medicine lay behind these techniques. For example, we now know that exposing tissue injuries, such as sprains, to cold reduces inflammation and decreases recovery time. Cold baths can be useful in controlling high fevers. Inhaling hot vapors helps congested sinuses

---

* Not to mention uncomfortable, given that I was with my mother.

and many respiratory problems. Sitting in hot and humid rooms is also known to help reduce joint and muscle pains. However, some animal-based research suggests that the benefits of heat, or, more specifically, the drawbacks associated with not being warm enough, go much further.

It should not come as a shock to anyone that experimentation on rodents is important to researchers trying to find ways to defeat cancer. The methods are simple enough. Scientists either inject tumor cells into rodents or use rodents that have been genetically engineered to develop cancer and then give the animals experimental drugs to see if they help fight off the disease.[38]

Drugs often appear useful in rodents but then prove ineffective in people.[39] This lengthy testing can be both costly and demoralizing. A better understanding of the differences between rodents and people would help, and this has led to some interesting experiments exploring temperature preferences and their effects on the immune system.

Mice in the wild* thrive at 86° Fahrenheit (30° Celsius). Because their small bodies shed heat quickly, warmth helps them avoid seeking out food as often in a predator-infested, or mousetrap-ridden, environment. Yet mice in laboratories are usually kept in environments that are somewhere between 68° and 79°F (20° and 26°C). This way lab technicians can clean the cages less frequently, as the mice drink and pee less often when it is cooler.[40]

Mice are perfectly capable of maintaining their body temperature in these environments. But researchers reporting in 2013 in *Proceedings of the National Academy of Sciences* discovered adverse effects.[41] A team led by immunologist Elizabeth Repasky at Roswell Park Cancer Institute in New York kept mice in enclosures at either 71.6°–73.4° or 86°–87.8°F (22.0°–23.0° or 30.0°–31.0°C). After giving the animals two weeks to

---

* Which is effectively the walls of my London apartment.

get used to the enclosures, the mice were injected with tumor cells from various cancer types and closely monitored.[42]

In the mice in the cool enclosures, the tumors grew at their normal, rapid pace; but in the mice kept in the warmer enclosures the tumors grew 50 percent slower.[43] Curious about whether the immune systems in the two groups of animals were functioning differently, Dr. Repasky looked at immune cells in blood samples collected from the mice. She found that cytotoxic T lymphocytes, cells that are particularly adept at finding and destroying cancer cells, were present in much higher numbers in the mice from the warmer enclosures than they were in the mice housed in the cooler enclosures.[44]

These discoveries hinted that warmth was somehow helping mice to more effectively fight off the tumor cells and led Dr. Repasky to wonder if mice ill with cancer might intentionally seek out warm places. To test this idea, she and her colleagues placed mice in specially designed enclosures with numerous temperature-regulated rooms. These rooms varied from a chilly 72°F (22°C) to a scorching 100°F (38°C). The healthy mice migrated to the 86°F (30°C) room to hang out. However, the mice with cancer migrated to the 100°F space and tended to stay there.[45]

The findings, Dr. Repasky argues, are not so much due to the healing benefits of heat as to the immune-system-suppressing effects of being cold. While a 71°F (21.7°C) laboratory might feel pleasantly warm to us, such a climate forces mice to divert resources in their bodies away from the immune system so they are better able to maintain their body temperature. Mice can survive in these temperatures, but are more vulnerable to disease. Thus, studying chilled mice for cancer is a flawed tactic.[46]

The human response to being cold is not much different. In modern society, where blankets, baths, and gas heating are everywhere, we don't see sick people traveling around the country in search of warm places to recover. But a thousand years ago the situation was very different. In a world where finding steady sources of warmth was hard, hot springs were nothing short of divine.

"And cold stress is only one among many stressors mediated through the sympathetic arm of the nervous system. There are other forms of stress, like psychological stress, that also seem to have negative immunological effects similar to those seen in cold-stressed mice," explained Dr. Repasky.

"This is exactly where our work is going next," explains Dr. Fredrickson. "We know that stress in general does not seem to have much of an effect on whether tumors appear or not, but it does seem to have a significant effect on how rapidly they progress once present."

So the temple at Bath was likely providing two real boosts to the immune system: the eudaimonic boost associated with collective worship, and the warm boost associated with soaking in the natural hot pools. Did people notice? Maybe. It is possible that so many people migrated to Bath simply because the temple was beautiful and because the springs felt good to soak in. However, I think it is equally possible that our ancestors picked up on the health benefits and explained them as the result of divine intervention without realizing something biological was at work.

## REGENERATION

Magical healing is not limited to the practices of the past. It runs deep in mythology too. The Greeks had the many-headed Hydra, which regrew its heads as soon as they were severed. Medieval stories are rich with tales of the green knight, an evil being who challenged King Arthur's men and could quickly regenerate both his head and his limbs when they were hacked off. Today, such miraculous healing abilities are tightly tied to superheroes, most notably Marvel's Wolverine.

Could our ancestors have encountered people or creatures with regenerative abilities? It seems unlikely. To date, no human or large animal has demonstrated the ability to spontaneously regrow severed limbs, though these tales were likely grounded, at least partially, in reality.

Regeneration is found throughout the natural world. Jellyfish and their kin do it. Earthworms, clams, and many flatworms do it too. An adult zebra fish can grow back both its fins and its heart. A salamander can regenerate its heart, jaw, arms, legs, and tail. So you have to wonder . . . were people just making things up, or were they inspired by real regenerative events in the environment? A couple of myths and a bit of science make me suspect the latter.

The ancient poet Hesiod tells us that the titan Prometheus was as much a lover of humanity as he was a troublemaker for the deities on Olympus.[47] First, he handed over the best meat from a slaughtered cow to mortals while passing along the gristle to the gods.[48] This greatly angered Zeus, who punished mankind by taking away the gift of fire. Keen to help humanity once more, Prometheus snuck into the halls of the gods and stole fire back. All of this ultimately led Zeus to both bind Prometheus to a mountain and order an eagle to come and rip out a chunk of his liver every day. While this would normally be fatal, the titan's liver regenerated overnight and the torture became eternal:

> *Prometheus he bound with inextricable bonds, cruel chains, and drove a shaft through his middle, and set on him a long-winged eagle, which used to eat his immortal liver; but by night the liver grew as much again every way as the long-winged bird devoured in the whole day.*[49]

Although the tale of Prometheus is well-known, another, similar Greek myth is not.[50] In this story, a giant named Tityos attempts to rape Leto, the mother of Apollo and Artemis.[51] The sibling gods shot their fiery arrows in anger at the giant, but his skin was too thick for them to cause harm. Forlorn, they turned to Zeus and petitioned him to take appropriate action. Convinced that Tityos had done wrong, Zeus had the giant chained in the underworld and arranged for two vultures to feast on his liver at every new moon. Like Prometheus, the giant's liver always grew back, making the punishment permanent. His torment is best described in *The Odyssey* when Odysseus sees him during a visit to Hades: "I saw Tityos, son of glorious Gaea, lying on the ground. Over nine roods he stretched, and two vultures sat, one on either side, and tore his liver, plunging their beaks into his bowels, nor could he beat them off with his hands." [52]

Both myths sound wild until you consider a study published in 1931 in the *Archives of Pathology*. That paper revealed that two-thirds of a rat liver could be surgically removed, and within seventy-two hours it would regenerate to its original size. [53] Humans, it turns out, are not much different.

"While the skin and the lining of the intestines can both regenerate, the liver is the only organ in our bodies that can completely regenerate itself even if most of it is removed," explains pathologist Dina Tiniakos at Newcastle University and the University of Athens.

However, the Greeks might have said it was the liver that was eaten and not some other organ, such as the pancreas, gallbladder, or heart, because the liver had long been viewed as the source of the soul. Even so, the scientist in me got to wondering whether this was just an intriguing coincidence or whether our ancestors were in some way aware of what human livers are capable of.

"A lot of us in the liver-research community suspect that the Greeks had to have seen rapid liver regrowth in animals and wounded soldiers and used their mythology to try and account for the phenomenon. Frankly, even now we kind of need the divine to explain this because we still don't really understand it," explains Dr. Tiniakos.

While liver regeneration is interesting, it is not as dramatic as the idea of limb regeneration. Given the sheer number of veterans who have lost arms and legs in combat, the gift of regrowth would be nothing short of a miracle. Yet it is a miracle that a research team led by Elly Tanaka at the Max Planck Institute of Molecular Cell Biology and Genetics in Dresden is striving to achieve.[54]

Using genetic engineering, Dr. Tanaka created salamanders in her lab that produced a jellyfish protein throughout their bodies. This protein was associated with bioluminescence, the process by which some animals glow in the dark. When the salamanders with this protein were exposed to blue light, they glowed bright green.

Dr. Tanaka and her colleagues collected bits of muscle, skin, and cartilage from the adult salamanders and transplanted these tissues into the limbs of ordinary salamanders. Once the transplanted tissues had stabilized and the animals had healed, the researchers started amputating limbs. This caused tissues at the severed regions to activate.

The glowing proteins were tethered to the regenerating tissues, so Dr. Tanaka was able to track them by shining blue light on the animals. To date, she and her colleagues have found that some tissues, such as muscle, are inflexible and remain muscle cells throughout regeneration. In contrast, some tissues, such as skin, can transform during regrowth into other sorts of tissues, such as cartilage. More important, this sort of work is beginning to reveal which tissues initiate and lead regeneration and which ones follow.[55]

Fibroblasts, which make up connective tissues in most animals, appear to be critical. "The fibroblasts are definitely the intelligent ones that seem to have the GPS system that tells other tissues what to do. There even appears to be a molecular circuitry between the fibroblasts that allows them to communicate with one another and coordinate tissue growth. It is really remarkable," explained Dr. Tanaka as I asked her about her latest work.

Whether such regenerative processes can eventually be induced in people depends upon what fibroblasts in the human body can be made to do. "We have fibroblasts all over our bodies. What we need to do is awaken them so they can go about the business of directing tissue regeneration," she told me.

Unquestionably important as Dr. Tanaka's work is, any limb regrowth that humanity sees using these methods will happen in a hospital under careful medical supervision. It is not quite the same thing as Hugh Jackman rapidly healing after being sliced and diced. Before setting out to write this book, I would never have dared to ask if we could create a mutant human capable of regeneration. But given the realities of Dr. Tanaka's work, I found myself wondering if regenerating mutant humans are possible.

## WEAPON X

Juvenile animals heal much more readily than adults, and embryonic animals are capable of even more impressive feats of healing than juveniles. Exactly why these differences exist has proved puzzling, but in 2012 a team led by stem-cell biologist George Daley at Harvard Medical School speculated that it might have something to do with a protein called Lin28.

"Little kids have this remarkable ability to heal rapidly that extinguishes as they grow," Dr. Daley explained during our first interview. Like most good interviews, ours wandered wildly and somehow ended up on the subject of the baby that my partner, Thalia, and I were expecting. "Listen, Matt, when your kid is born, you are going to be amazed. One week you will forget to clip her nails and then you are going to find scratches all over her face. You'll panic initially, but then in, like, five hours the scratches will be gone. Our team knew that this accelerated healing vanished somewhere in early infancy, and we wanted to know why."

Dr. Daley knew Lin28 played a role in development and that if Lin28 was present in young worms in high doses, it delayed the onset of adulthood. Similarly, if Lin28 was actively removed from juveniles, it triggered sudden progression into adulthood. He also knew that Lin28a, a gene found in mammals that is closely related to Lin28, was tightly linked to programs of growth, development, and metabolism, and that its presence declined as mammals aged. This made him wonder if it was connected to the rapid healing seen in infants. Keen to explore this, he and his colleagues created mutant mice in the lab that, rather than only express Lin28a as embryos, expressed it throughout their lives. The resulting mutants grew to be huge, hairy, and, most important, capable of remarkable feats of regeneration.

Dr. Daley and his colleagues reported in the journal *Cell* in 2013 that when the ears of the mice were intentionally wounded, they rapidly healed without scars.[56] During routine tagging procedures, the mice, as juveniles, had the tips of their toes cut off. Normally, such serious wounds leave scars and the toes do not grow back, but in the mutant mice the toes fully regenerated.

Dr. Daley argues that these healing abilities appeared because the raised Lin28a levels in juveniles and adults led tissues to behave as if they were still in an embryonic state, where rapid healing would normally take place.

Yeah, I know what you're wondering: Would creating a mutant human with raised Lin28a levels give us a rapidly regenerating mutant like Wolverine? Well, you're not alone. When this research was reported in the online news section of the journal *Nature*, one of the editors there stuck a photograph of Hugh Jackman with claws sprouting from his hands as the main image for the article, and lots of us in the science-journalism world were half joking over Twitter that the work was well on its way to making X-Men a reality.

"I've got to tell you, before we published our paper, I didn't even know who Wolverine was! When he started getting mentioned by the press, I had to ask my kids what people were talking about," Dr. Daley said, laughing.

I laughed too but found myself daunted by what I had to ask: "I know we live in a world where we can't go around mutating babies for the sake of science, but if we could test your Lin28a work on humans, would we make Wolverine a reality?" I expected a long pause to follow such an audacious question.

"I think we would see much the same thing as we see in the mice: big and hairy people who heal really rapidly," responded Dr. Daley without skipping a beat.

But would they tolerate having adamantium fused to their skeletons?* Seriously, I wanted to ask that one, but Dr. Daley cut me off before I could thoroughly embarrass myself.

"Here's the thing, though: the problem with Lin28a is that it is a double-edged sword. Sure, it grants rapid healing, but it also substantially increases the chances of someone developing certain types of cancers." The trick, Dr. Daley suspects, is going to be exposing people to Lin28a in short bursts to help specific organs regenerate when such regeneration is desperately needed. The kidneys are currently a key target. "We know that if you expose kidneys to Lin28a over a long period of time, they initially grow quite a lot but later develop tumors. We're thinking that if you take a patient suffering from kidney failure and therapeutically raise their Lin28a levels, this might allow kidney regeneration and help the patient avoid dialysis treatments."

It is amazing stuff and I can't wait to see it put to use. Nonetheless, creating a regenerating mutant is more about altering the human body in remarkable ways than it is about healing magic. So, I think it's time we bid farewell to curative rituals and explore the myths and realities associated with supernatural transformations.

---

* Okay, for those not steeped in comic book lore, as part of a government program called Weapon X, which was designed to use mutants in war, Wolverine had the fictional metal adamantium fused to his bones during a surgical procedure that he survived only on account of his regenerative abilities.

# TRANSFORMATION

## Into Animals, Berserk Warriors, Supersoldiers, and the Other Gender

*You know, I hate to say "I told you so,"*
*but that supersoldier project was put on ice for a reason.*

—TONY STARK, *THE INCREDIBLE HULK*

Wolverine's name stems from his quick temper and claws. He is not actually part animal. However, older mythology is rich with tales of magic that does transform men into beasts. Among the most famous of these myths is in Homer's *Odyssey*.[1]

During their wanderings, Odysseus and his crew arrive at the island of Aeaea. Starving and exhausted, they split into two groups to search for resources, and one group stumbles upon a palatial home. A beautiful woman emerges and welcomes the men inside for a feast. All but one of them follow her.

This is no ordinary woman. This is the sorceress Circe, fabled daughter of the goddess of magic, Hecate, and as she prepared the feast for the men, the story says she

*made for them a potion of cheese and barley meal and yellow honey with Pramnian wine; but in the food she mixed baneful drugs, that they might utterly forget their native land. Now when she had given them the potion,*

*and they had drunk it off, then she presently smote them with her wand and*
*penned them in the sties.*[2]

Terrified by what he sees, the one crew member who did not enter Circe's home races into the wilderness to tell Odysseus of the horrible incident. Determined to rescue his men from this witchcraft, Odysseus travels through a forest toward Circe's home and, along the way, encounters a young man who reveals himself to be Hermes, a god most often thought of as a messenger deity but who was often associated with healing. Hermes, aware of the danger, digs up a plant in a nearby meadow and hands it to the hero, saying:

> *Here, take this potent herb, and go to the house of Circe, and it shall ward*
> *off from thy head the evil day. And I will tell thee all the baneful wiles of*
> *Circe. She will mix thee a potion, and cast drugs into the food; but even so*
> *she shall not be able to bewitch thee, for the potent herb that I shall give thee*
> *will not suffer it.*[3]

What is most intriguing about the text is that it is specific about where Hermes's medicine comes from and what it looks like. Odysseus says:

> *So saying, [Hermes] gave me the herb, drawing it from the ground, and*
> *showed me its nature. At the root it was black, but its flower was like milk.*
> *Moly the gods call it, and it is hard for mortal men to dig; but with the gods*
> *all things are possible.*[4]

True to the god's word, the moly renders Odysseus invulnerable to Circe's magic. He drinks her potion, eats her food, and stands there as she points her wand at him. However, when she tells him to join his friends in the pigsty, he resists the spell and rushes her with his sword.

Without the mention of drugs being mixed into the food, it would be easy to shrug off such a myth as the stuff of pure imagination. However, that Circe's power seems to be stemming from a knowledge of poisons suggests something real might be behind the tale.

One of the most common arguments is that Circe was feeding the crew jimsonweed. While that sounds innocent enough, jimsonweed—or *Datura stramonium*, as it is known in the scientific world—belongs to the deadly nightshade family and contains high levels of anticholinergic alkaloids such as scopolamine, hyoscyamine, and atropine.[5] These compounds block the vital neurotransmitter acetylcholine from interacting with its receptors in the brain. When this neurotransmitter is blocked, we can't distinguish reality from fantasy, we exhibit bizarre behavior, and we can suffer pronounced amnesia.[6] "Patients who consume this stuff often have vivid hallucinations and become seriously delirious," advised Harvard Medical School toxicologist Alan Woolf. Really, I think one academic analysis of the anticholinergic alkaloid cocktail sums things up rather nicely: "No other substance has received as many 'train wreck' negative reports [with] the overwhelming majority finding their experiences extremely mentally and physically unpleasant and not infrequently physically dangerous."[7]

We think she was using *Datura* and not some other poison for two reasons. First, jimsonweed is found all over the classical world.[8] Second, *The Odyssey* makes it clear that Circe expects the crew to forget their fatherland. More specifically, she is expecting an amnesia-like effect on par with what *Datura* actually causes.[9]

As for her use of the wand, the earliest pottery depicting Circe does not show her pointing the wand at the men but rather mixing poison into food with it.[10] Given this art, a case can be made that the wand in the original story was not a wand at all but an ordinary stirrer that ultimately came to be drawn as a magical object rather than an everyday kitchen utensil. Or perhaps the wand was just a theatrical prop to punctuate Circe's power.

TOP: The side of a drinking cup showing the origins of the magic wand in Circe's hands as a stirrer for poison. Greek, Archaic Period, about 560–550 BC. © 2015 Museum of Fine Arts, Boston (detail).

BOTTOM: Ulysses at the Palace of Circe. The wand's history as a stirrer is all but forgotten. Wilhelm Schubert van Ehrenberg, 1667, oil on canvas. © The J. Paul Getty Museum, Los Angeles, Gift of Mr. and Mrs. Thomas Brant (detail).

I find it fascinating that, over the centuries, this role of the wand in magic has not changed much. In one of the core texts that many modern stage magicians read while in training, Professor Hoffmann's *Modern Magic*, the wand is described as an ideal prop for distracting the eyes, for granting an excuse to close the hand and hide something within it, and to provide the illusion of having real magical power.[11]

Intriguingly, this aspect of the Circe tale lives on in modern mythology. Her use of *Datura* and the wand is almost identical to the "magic" wielded by the *Batman Begins* villain Dr. Jonathan Crane, aka Scare-

crow. He sprays poisonous gas at his opponents and then throws a bur-
lap mask over his face to create the terrifying hallucination that he is
being eaten alive by maggots so his targets become mentally unhinged.
Yet *Datura* is not the only real-world element found in *The Odyssey*.
Remember how Sauron's ring "fell out of all knowledge" as it sat for centuries in Gollum's hands beneath the Misty Mountains? This is similar
to what happened with the protective plant that Hermes gave Odysseus.

In 1951, during the Cold War, pharmacologist Mikhail Mashkovsky
at the Russian Academy of Science discovered that villagers in the Ural
Mountains were rubbing ground-up snowdrop flowers into their skin as a
pain reliever and using the snowdrop plant as a medicine to help children
afflicted with polio to fight off the paralysis that was so often caused by
the dreaded disease.[12] Fascinated by this, Dr. Mashkovsky and colleague
Rita Kruglikova-Lvova explored the effects of a compound extracted
from snowdrops on the muscles of frogs, rabbits, and guinea pigs and
the brains of rats. This experimentation led the researchers to conclude in
the late 1950s that the compound had the ability to protect specific neurotransmitters from being damaged by diseases and toxins.[13] The snowdrop compound galantamine was soon registered as Nivalin in Bulgaria in
1959, and from there it began its slow journey into the West.[14] Today, the
drug is proving more valuable than ever because of Alzheimer's disease.[15]

First described in 1907 by the German psychiatrist Alois Alzheimer,
the disease is vicious in the manner by which it progressively destroys
memory and cognitive function. In the late 1970s, a team of researchers
found that people who had died of Alzheimer's commonly had brains
deficient in acetylcholine.[16] This discovery paired well with reports that
the most common brain deficit found in Alzheimer's patients was cholinergic in nature and led several research groups to argue that the disease
arose from an inability to transmit signals across cholinergic synapses.[17]
Galantamine's ability to protect acetylcholine seemed perfect for staving
off the effects of Alzheimer's, and it is now widely used for this purpose.[18]

Could this be Homer's moly? In 1981, as the drug's origins started to become better known, neurologists Andreas Plaitakis at the Mount Sinai School of Medicine in New York and Roger Duvoisin at Rutgers Medical School in New Jersey proposed at the Twelfth World Congress of Neurology that snowdrop might have been the plant that Hermes handed to Odysseus.[19] To support their argument, they pointed out that the plant was commonly found in Greece, that it grows in forest glens like the one visited by Hermes, and that it is an effective antidote to *Datura*. They also noted that its petals were milky white and that it had a darkly pigmented root just as the moly described in *The Odyssey*. Moreover, galantamine is not like many other anticholinesterases that break apart and become useless in the body rather quickly. Galantamine endures, producing a lasting protection that prevents acetylcholine from being blocked, which would have made it perfect for the situation Odysseus found himself in.[20]

Did the Greeks really know that snowdrop could stave off neurotransmitter-attacking poisons and diseases? I suspect they did. There isn't a lot of direct evidence, but Theophrastus, a Greek writer from the fourth century BC, writes in his text *Historia plantarum* that moly "is used as an antidote against poisons."[21] This knowledge must have been woven into stories long ago and transformed from history into legend and from legend into myth in most parts of the world. What is wonderful is that the myth lives on.

Recently, after a long day of writing, I ambled into the living room and found Thalia equally exhausted after work and watching the film *Stardust*. As I sat next to her on the sofa sipping my tea, I watched as the evil witch in the tale tries, and fails, to harm the hero, Tristan, with her dark magic. The reason for her failure? He was wearing a milky-white flower on his coat. The species looked familiar, so I pulled up the script of the film the next morning, and sure enough there was its name:

Girl: You shouldn't buy the bluebells. Buy this one instead. Snowdrop. It'll bring you luck. [22]

If snowdrop was Homer's moly, over the ages it was almost forgotten. For the sake of those suffering from Alzheimer's, I'm glad Dr. Mashkovsky found its powers before they were completely lost to the ravages of time.

Does this mean that Odysseus and Circe were real? In the literal sense, I doubt it. However, a talented female poisoner might have been living on an island who used her knowledge to lead natives to worship her as a demigod. At some point in history someone might also have learned that consuming snowdrop provided protection against certain diseases and poisons. A local hero might have known a thing or two about herbalism, stood up to a cruel poison-wielding witch, proved that he was resistant to her magic, and come to be known as the great-grandson of Hermes . . . and wouldn't that be cool?

## GOING BERSERK

Not all historical animal transformations were curses. Some were blessings. Among the peoples who once dwelled in what is now Norway, Sweden, and Finland, certain men, during the heat of battle, were said to transform into beasts and attack their enemies in an unstoppable frenzy.[23] Ancient reports note that their faces became hideously altered, and they hissed, bared their teeth, and howled like wolves. Woe to anyone who had to do battle with them, for the *Ynglinga Saga*, an old Norse saga written by the poet Snorri Sturluson around AD 1225, states that they "went [into battle] without mailcoats [armor] and were as wild as wolves, they bit their shields, were as strong as bears; they killed people, but they themselves were hurt by neither fire nor iron."[24]

Their intense battle fury was known as *berserkergang*, and in contrast to what Circe did to Odysseus's crew, the *berserkergang* transformation was welcomed. The Vikings believed it was granted to them by the power of their god Odin during times of war. This makes a lot of sense, since becoming a raging and nearly invulnerable warrior on the field of battle would be a tremendous advantage. Could it possibly be real?

In 1784 the Swedish scientist Samuel Ödman suggested that the fungus *Amanita muscaria*, which is best described as the red-and-white mushroom from the video game Super Mario Bros.,* was being consumed to trigger this transformation.[25] In 1886, the Norwegian botanist Fredrik Schübeler further supported this argument. So too did the Swedish scientist Hugo Hildebrandsson in 1918 when he reported to the Royal Scientific Society in Uppsala that during the war that took place between Sweden and Norway a hundred years earlier, some soldiers consumed *Amanita muscaria* and then entered a wild rage on the battlefield.[26] However, in 1929 a Norwegian medical historian named Fredrick Grön challenged these theories for being based upon thin evidence, and so too have many others since.[27] Most of the arguments supporting the use of this mushroom by the berserkers are based upon mythology and symbolism.

In *The Gylfaginning*, one of the earliest written versions of the Norse creation mythology, where the divine land of Asgard is described, the all-powerful god Odin is said to sit on high where he can scan the entire landscape with his eye and know all.[28] You may note that I say *eye* rather than *eyes*, and this is because Odin traded one of his eyes early in life. Mimir, the owner of the well of wisdom found beneath the World Tree, was rumored to be supernaturally sensible because he drank from his magical well every day. Odin was keen to gain this sort of wisdom, and to obtain it he pledged one of his eyes in return for a single drink. Mimir consented, and Odin drank his fill.[29]

It all sounds innocent enough, but mythologists have found another story, this one from the Mordvin people, who now inhabit much of Russia, which hints at a fungal origin for Odin and Mimir. The story tells of a giant birch tree growing in the heart of a forest. The great tree's roots spread out around the world, and its branches reach into the

---

* Yes, Mario and Luigi were taking mushrooms to grow to giant size. They were berserker, fireball-throwing Italian plumbers.

heavens. At the base of this mighty birch is a spring, roofed with white sheets. A red wooden bowl filled with a sweet drink sits by its edge. A silver ladle decorated with stars accompanies the bowl.[30]

Mythologists have analyzed this scene in detail. Some argue that the red bowl is representative of the color red found on *Amanita muscaria*. Some argue that the silver ladle and the mention of stars are descriptors of the mushroom's white stem and spots, which look vaguely silver in moonlight.[31] One element of this analysis that I buy into is the mention of the birch tree. *A. muscaria* tends to grow under birch trees and usually develops close biochemical associations with them. Odin's nickname Sithhott means "the long hooded," and Sidhhottr means "he of the broad hat." Both have been interpreted to mean Odin himself is symbolic of a mushroom. Personally, I'm skeptical of these interpretations. Maybe the Vikings were associating Odin with the mushroom, maybe they weren't. What I think we can say is that the powers of the berserker warriors clash with the biochemical effects that *A. muscaria* has on the human body.

According to a review on the biochemistry of *A. muscaria*, published in the journal *Mycological Research* in 2003, eating one of these mushrooms can cause a wide array of unpleasant symptoms, most commonly including confusion, dizziness, nausea, and exhaustion.[32] People consuming these mushrooms also typically become hypersensitive to visual, olfactory, and auditory stimuli and sense space and time as distorted.[33] Indeed, these distortion effects have led several scholars to propose that when Lewis Carroll wrote *Alice's Adventures in Wonderland*, he had *A. muscaria* in mind when Alice grew and shrank after consuming parts of a mushroom.[34]

I found it hard to believe that any Viking warrior could effectively fight while suffering from dizziness, exhaustion, and vomiting, and when I ran this past Donald Pfister, a mycologist at Harvard University, he expressed his doubts too: "I've always been skeptical about these fungi having hallucinogenic properties or about them granting any sort of powers. It is possible that there are some species in Scandinavia with different effects, but I doubt it. Honestly, feedback that I've had from folks who tried these has never suggested that they did much for them." This led me to wonder if the berserkers might have been doing something more with *A. muscaria*. Some researchers suggest that they were consuming the mushrooms with vast amounts of wine or mead.[35] Might getting high on *A. Muscaria* while drunk generate the legendary berserker effects? Maybe, but Dr. Pfister was quick to remind me that such combinations were quite dangerous too: "A lot of fungi can make you deathly ill when mixed with alcohol." When I looked to Dr. Woolf for further toxicological advice, he emphasized how unpleasant they are to take. "No one tries to take *Amanita muscaria* mushrooms on purpose. They generally cause a very nasty experience, including staggering, twitching, abdominal cramping, and sometimes comas," he explained. With Dr. Pfister's and Dr. Woolf's comments in mind, I turned my attention to options presented by the field of archaeology.

In 1977 an archaeology team discovered several hundred seeds from the plant *Hyocyamus niger* in a Viking grave near Fyrkat, Denmark.[36] Better known as henbane, *H. niger* is a serious chemical cocktail with each of its various parts having different effects.[37] If crushed and rubbed on the skin, the seeds numb pain.[38] Thus the Greeks used it to dull toothaches.[39] If eaten, both the seeds and the leaves cause hallucinations.[40] When the flowers are boiled, crushed into an ointment, and massaged into the body, people feel as if they are flying while also experiencing mild visual hallucinations.[41] While using too much of any part of the plant or preparing it improperly can kill or induce a seemingly supernatural slumber, I think descriptions of its effects match the berserkers' behavior well enough that it could have been behind their legendary rage.[42] "Much like *Datura*, it is an anticholinergic. So we'd expect delusions, hallucinations, and agitated behavior from people taking it," explained Dr. Woolf. Henbane has only been found at one Viking site so far, and a sample size of one doesn't stand for much.[43] With that said, henbane seeds are tiny, and all the other bits of the plant rot away quickly, so it may have been widely used yet not left much evidence for archaeologists to find. I believe the berserkers were taking something real that was granting them their powers, but if we want to get a better idea of what that was, I think the archaeologists need to keep digging.

## TOWARD THE FIRST AVENGER

While it can be debated whether and how the transformations found in Homer's *Odyssey* and the Norse legends actually happened, nobody would dare to argue that such transformations are only the stuff of fiction. Drugs abound that have the potential to alter both our physical performance and our bodies. Almost all of my work on this topic, as a science journalist, has explored the arms race between professional organizations trying to keep competitions fair and the athletes trying to cheat by taking drugs that enhance their abilities. But ethics and laws aside, could I load myself up on a regimen of drugs and transform into a supersoldier?

The best place to start with such a question is amphetamines. These drugs increase the release of the neurotransmitter dopamine, which helps us to stay alert, and norepinephrine, which helps us to concentrate. They also inhibit reuptake of these neurotransmitters and thus stimulate the brain overall. Given their effects, it should not come as a surprise that amphetamines are often prescribed to treat people whose maladies, such as attention deficit disorder, prevent them from focusing on their work, but the drugs also boost mental performance in people without such disorders.

Improved concentration and alertness are definitely useful abilities for a supersoldier. Indeed, amphetamines, which were first synthesized during the 1920s, were handed out to troops during the Second World War for precisely this reason. However, the benefit of doping people up on these drugs goes far beyond improving mental status.

With enhanced dopamine and norepinephrine production, our physical performance is improved. We know this because in 1959 two researchers, Gene Smith and Henry Beecher at Harvard University, used fifty-seven competitive swimmers, runners, and shot-putters as guinea pigs in an experiment to test the powers of amphetamines.[44] Half of the subjects received amphetamines; the other half, a sugar placebo. The participants were then put through a total of eight hundred events during mock competitions that were meant to feel like the real thing. Athletes swam, ran, and threw weights up to 4 percent better when they were drugged than when they were not. Reaction time, strength, and acceleration are improved with amphetamines, but the drug is especially effective against exhaustion. Amphetamines dramatically delay fatigue during hard labor or strenuous exercise. So, amphetamines seem like sensible drugs for a supersoldier to be taking—but what else?

During the 1950s and 1960s, to help doctors fight the tissue-wasting effects of certain infectious diseases, anabolic androgenic steroids were produced.[45] These were a godsend for patients, because they replicated the effects of testosterone and other hormones in the body that often played an important role in reconstructing damaged tissues. However, people quickly figured out that these steroids could be useful outside of medicine too.

In a healthy person who is regularly working out, tiny tears appear in muscles when they are stressed. This is particularly true when people, such as weight lifters, push specific muscles to their limit. These tears heal up during rest periods between workouts as muscle tissue replicates around the damaged areas. Through this natural process muscles increase in size over time, but it can be made into quite an unnatural one with anabolic androgenic steroids.

Steroids promote muscle growth and make people, particularly men, a lot stronger. Moreover, most of the muscle increase is found in the upper arms and shoulders. This is perfect for throwing more powerful punches, bending bars, lifting heavy gates, and all the other sorts of things that supersoldiers are supposed to be able to do. And steroids decrease recovery time. So these definitely sound important for any aspiring supersoldier.

You rarely see Captain America breathing hard. It doesn't matter if he just ran after an aircraft and grabbed hold of its wing or beat up twenty goons trying to ambush him in an elevator. Nothing seems to wind supersoldiers, and I wondered if any real-world compounds could grant such a boon. As it turns out, some can.

The hormone erythropoietin helps the body make better use of oxygen.[46] This hormone, which occurs naturally in all of our bodies, plays an important part in creating red blood cells, which are crucial for transporting oxygen.[47] The more of them we have, the harder we can push ourselves before feeling that we aren't getting enough air.

Artificial erythropoietin started being made in labs during the 1980s to help patients with anemia generate more red blood cells.[48] While the blood of most normal people consists of around 42 percent red blood cells, erythropoietin supplements can push those levels to 52 percent and sometimes over 60 percent. This improves oxygen transport and allows each breath and pump of the heart to provide tissues with more of the vital resources that they need.

On top of everything else, a supersoldier needs courage. Trembling in the face of serious danger is no good, particularly if you are trying to aim a weapon at enemies trying to kill you. If I'd trained at West Point and fought my way through Afghanistan, I might not need to take anything to get control of my anxiety. That's not me, though, and I know that if I had to face off against villains with missiles or, worse, superpowers at their disposal, I'd be running away* in a panic.

In Britain, liquid courage is most commonly dispensed in pubs, and while it might seem inappropriate for a supersoldier to have a few pints before battle, it isn't that ridiculous. The first athlete to ever be caught doping by the International Olympic Committee was a Swede, a pentathlete, at the 1968 Olympics in Mexico City.[49] In the pentath-

---

* And multitasking by soiling myself at the same time.

lon, being able to shoot a rifle with high accuracy after lots of physical exertion is important. Ethanol, a depressant found in alcoholic beverages, was discovered in the Swede's urine, and this got him disqualified. Why? Ethanol reduces the pulse and diminishes tremors that make it harder to shoot a gun with accuracy. So I could add a shot of vodka to my supersoldier recipe, or I could look for something better . . . such as a beta-blocker.

Beta-receptors are parts of the nervous system that get kicked into action by certain neurotransmitters, such as norepinephrine. When this happens, body temperature goes up, the skin starts to sweat, the eyes dilate, and anxiety follows. Lots of beta-receptor activity makes us tremble. Drugs that block beta-receptors from getting activated, known as beta-blockers, were first developed to help calm down cardiac activity during heart attacks, but they've since been abused by athletes competing in gymnastics, golf, archery, and shooting, where steady hands are essential.[50] Crucially, beta-blockers seem to be the most helpful to less experienced athletes. Thus, for a supersoldier who has never seen any sort of active combat before, taking beta-blockers would be valuable.

So, to review, if I took a mix of amphetamines, anabolic androgenic steroids, erythropoietin, and beta-blockers, I'd have faster reaction times, better endurance, increased strength, improved respiratory function, and nerves of steel. You might think gaining all of this might require me to take dozens of pills and stab myself with lots of syringes every day, and while that would once have been the case, it isn't anymore.

Most people think of viruses as pathogens that kill, but some are now being grown in labs and modified so they can carry compounds to specific places in the body. These viruses are most often used like smart bombs, carrying medications to diseased tissues that need treatment, but they can also be used in more creative ways. For example, performance-enhancing genes can be inserted into the body through viruses.[51] Using additional genes in this way has the potential to alter a living person's genetic features. This means that genes for increased erythropoietin and muscle-building hormone production could be ferried into my body by viruses and added to my DNA, permanently making me both stronger and a more efficient breather. Thus, it is theoretically possible to put me through a series of laboratory treatments that would permanently transform my body. Just give me a bulletproof shield and a patriotic spandex outfit and I'll be all set to fight for truth, justice, and the American way! Now for the fine print . . .

These drugs have side effects. While some are relatively mild—amphetamines give people upset stomachs, and anabolic androgenic steroids tend to cause acne—some are serious.

"A lot of people forget that the heart is a muscle. So when they take steroids, the heart gets bigger and thicker than it naturally should, and this leads to heart disease," explains Olivier Rabin, director of science at the World Anti-Doping Agency. As if that were not bad enough, steroids can cause strokes, liver dysfunction, cancer, and testicular atrophy. Erythropoietin is hugely problematic as well. By increasing red blood cells, it makes the blood thicker, which can clog the capillaries in our brains and cause strokes. "Our heart was simply not built to pump jam," says Dr. Rabin. As for amphetamines, they cause heart problems and seizures, and are likely associated with a host of mental side effects as well.

Richard Friedman, a professor of psychiatry and director of the psychopharmacology clinic at Weill Cornell Medical College, proposed in a 2012 opinion piece in the *New York Times* that the consumption of amphetamines by American troops in Iraq and Afghanistan is likely causing serious psychological problems.[52]

While just 0.2 percent of the troops returning from tours of duty in 2002 were suffering from post-traumatic stress disorder, this number leapt to 22 percent in 2008.[53] Could the length and nature of these two military campaigns account for the difference? Sure, but Dr. Friedman found that prescriptions of amphetamines to soldiers jumped from three thousand to thirty-two thousand during these years too.[54] Amphetamines support the building of memories, particularly during moments of high emotional intensity.[55] He argued that these drugs induce post-traumatic stress disorder by amplifying memories of the horrors of war.[56]

Do all of these drawbacks make the idea of an engineered Captain America impossible? No. With the right mix of drugs and therapies, soldiers with remarkable abilities can be created and sent off to fight. The trick is going to be keeping these transformed warriors from succumbing to heart disease, cancer, and madness.*

Warfare, athletic cheating, and sarcastic superhero speculation aside, our ability to dramatically alter the human body has the potential to do a great deal of good in other areas. Indeed, it is already saving lives.

## GENDER BENDER

I remember a fair few moments as a teenager while I was sitting next to my parents when the topic of sex came up on television or during a film. It was uncomfortable to be sure, but over the years it mostly faded. I say *mostly* because there was this time when I was twenty-four . . .

My father decided in midlife to become a cycle maniac. He started going on these crazy rides to raise money for charities. He cycled from San Francisco to Los Angeles, from Anchorage to Fairbanks, and eventually from Amsterdam to Paris. I was already living in London by then, so it was easy enough to take the Eurostar over to France, catch up with my mother for a day, and then meet Dad at the finish line. Being an art historian, my mother had only one activity in mind for our free time: the Louvre.

Now, most kids rebel against their parents by staying out beyond curfew, driving too fast, and drinking while underage. That was never me. Mom was a scholar of the Italian Renaissance, and so, in teenage rebellion, I decided I would detest the Renaissance and develop

---

* And, yes, keeping their testicles attached to their bodies too.

a passion for the Dutch golden age. Whenever we ended up in an art museum, the debate continued. Thus it was so with the Louvre.

Tired of looking at the likes of the *Mona Lisa* and *St. John the Baptist* in the Richelieu wing, I insisted we migrate to the Dutch gallery, and eventually Mom relented. The Louvre is something of a labyrinth, and stumbling upon weird art that you never knew existed is part of the experience. During such wanderings we came upon a hall filled with marble sculptures from the 1600s.

The style was baroque and thus neutral ground. Several pieces were by Bernini, a masterful sculptor who made marble come to life as almost nobody else ever could. One sculpture of a woman lying on her side beckoned us closer. She was gorgeous. Well-proportioned hips, smooth skin, and a perfect back. The sheets on her legs looked as if they had been made of pure Egyptian cotton. As we walked around the sculpture, her beautiful breasts came into view, and then her penis. We stopped dead in our tracks and stood there in silence for a few awkward moments. The sign next to the statue was simple enough: SLEEPING HERMAPH-RODITUS. Mom cleared her throat and, after a few moments, explained, "These were really popular with the Romans and Greeks. They made a lot of them." More silence. "Shall we, uh, continue on to the Rembrandt room?" Mom's *wanting* to go see Dutch art—that spoke volumes. We did move on, but that penis never left our minds.

Hermaphroditus was the son of Hermes and the goddess of love, Aphrodite. According to the poetry found in Ovid's *Metamorphoses*, published in AD 8, he was an attractive young man, and during his wanderings he caught the attention of the water nymph Salmacis.[57] The poem tells how she fell for him at first sight and sought to bring Hermaphroditus close for an embrace. However, Hermaphroditus felt nothing for the nymph and told her to leave him alone. Saddened, she departed, but kept a close eye on him from the forest as he continued on his journey along the water's edge.

*Playful and wanton to the stream he trips,*
*And dips his foot, and shivers as he dips.*
*The coolness pleas'd him, and with eager haste*
*His airy garments on the banks he cast;*
*His godlike features, and his heav'nly hue,*
*And all his beauties were expos'd to view.*
*His naked limbs the nymph with rapture spies,*
*While hotter passions in her bosom rise,*
*Flush in her cheeks, and sparkle in her eyes.*
*She longs, she burns to clasp him in her arms,*
*And looks, and sighs, and kindles at his charms.*

*Now all undrest upon the banks he stood,*
*And clapt his sides, and leapt into the flood:*
*His lovely limbs the silver waves divide,*
*His limbs appear more lovely through the tide;*
*As lillies shut within a crystal case,*
*Receive a glossy lustre from the glass.*
*He's mine, he's all my own, the Naiad cries,*
*And flings off all, and after him she flies.*
*And now she fastens on him as he swims,*
*And holds him close, and wraps about his limbs.*
*The more the boy resisted, and was coy,*
*The more she clipt, and kist the struggling boy.*
*So when the wrigling snake is snatcht on high*
*In Eagle's claws, and hisses in the sky,*
*Around the foe his twirling tail he flings,*
*And twists her legs, and writhes about her wings.*

*The restless boy still obstinately strove*
*To free himself, and still refus'd her love.*

*Amidst his limbs she kept her limbs intwin'd,*
*"And why, coy youth," she cries, "why thus unkind!*
*Oh may the Gods thus keep us ever join'd!*
*Oh may we never, never part again!"*

*So pray'd the nymph, nor did she pray in vain:*
*For now she finds him, as his limbs she prest,*
*Grow nearer still, and nearer to her breast;*
*'Till, piercing each the other's flesh, they run*
*Together, and incorporate in one:*
*Last in one face are both their faces join'd,*
*As when the stock and grafted twig combin'd*
*Shoot up the same, and wear a common rind:*
*Both bodies in a single body mix,*
*A single body with a double sex.*

*The boy, thus lost in woman, now survey'd*
*The river's guilty stream, and thus he pray'd.*
*He pray'd, but wonder'd at his softer tone,*
*Surpriz'd to hear a voice but half his own.*
*You parent-Gods, whose heav'nly names I bear,*
*Hear your Hermaphrodite, and grant my pray'r;*
*Oh grant, that whomsoe'er these streams contain,*
*If man he enter'd, he may rise again*
*Supple, unsinew'd, and but half a man!* [58]

Salmacis's wish that she and Hermaphroditus could become one was answered, and Hermaphroditus was transformed.

The story is the stuff of imagination, but the idea of an individual of mixed sex is very real. During conception, sometimes two eggs are present rather than one. Both eggs can be fertilized at the same time by different sperm with one carrying a female X chromosome and one carrying a male Y chromosome. Such circumstances usually lead to fraternal twins of different sexes if both eggs attach in the uterus. But the two eggs occasionally fuse, forming a mixed-sex baby.

Another frequent pathway toward mixed sex occurs during cell division, when a sperm's sex-determining region migrates from the Y chromosome to the X chromosome and remains there. Again, a child formed from a sperm carrying such information will be born with mixed sex.

What does mixed sex look like? It depends. In some cases, an individual who looks much like Bernini's *Sleeping Hermaphroditus* will be born. In others, the physical changes are more subtle or barely visible.

The biological processes responsible for causing mixed sex are not new. Our bodies have been behaving in the same way for millennia, and we must accept that people with mixed sex have always been part of our population. Society's predilection for putting everyone into tidy categories* makes it hard to know just how many mixed-sex people are alive today, and even harder to work out how many were alive long ago, but biology tells us that mixed-sex people have always been a part of the global fabric.

Was Ovid's poem based upon observations of a real person of mixed sex who lived during his time? As Ovid is dead and gone, we will never know, but we do know that modern medicine is rapidly making this mythical magic into reality for people with another condition known as gender identity disorder.

For reasons that are still not well understood, from a young age, some children are born identifying with a gender that does not match with their physical sex. They often have a dislike of their genitalia and experience a great deal of stress when society forces them to behave in

---

* Australia has commendably created a third category for gender on its passports. In addition to the traditional *M* and *F*, there is now also the *X* category, for intersex individuals.

the role of a gender that they do not identify with. As they enter puberty and their bodies fully mature into a form that their minds resent, depression and suicide are all too common.

In the interest of reducing misery and saving lives, numerous research groups around the world have explored using drug therapies to help the bodies of these patients transform during puberty into the bodies that their brains tell them they ought to have. The biochemistry of it is relatively straightforward.

First, drugs that mimic the hormone gonadotropin are administered. These drugs find the receptors where gonadotropin would normally be released and interfere with them.[59] This treatment is biologically vital as it shuts down sexual changes that would naturally start at puberty and is particularly valuable as it is reversible so long as it is not combined with any other therapies.

Gonadotropin-releasing hormone analogs only shut down puberty so long as they are taken.[60] If they stop being taken, puberty resumes. This means that young people experiencing gender-identity challenges, along with all the other crap that goes on in school during those years, are given the gift of time by taking them. If, after a year or two, they and their supportive doctors think it best to end the treatment and let the body develop as it normally would, that is easy enough to do. If instead someone decides to take the next step and begin physical transformation into the opposite sex, that is not too hard either. What is hard is helping society to understand that these treatments are safe and that gender-identity confusion stems from biology rather than psychology.

As I pored over the papers on all of this work, I realized that my training in paleontology put me in no position to properly grapple with the subject, so I did what any self-respecting science journalist would do and arranged to have coffee with an expert. Most of the top people in this field are in the Netherlands, but since I was visiting Boston at the time, I decided to track down Norman Spack at Harvard Medical School, the man who brought the Dutch techniques to the United States.

"There's all this concern about whether gonadotropin-releasing hormone analogs are going to mess up child brain development, but it's ridiculous, because we've been using them for decades and the kids on them have grown up just fine," he explained.

These drugs have long been administered to treat another condition, known as precocious puberty. In this disorder, kids start growing up too early, taking on their adult forms at ages as young as four. It isn't physically good for the body and leads to all sorts of psychological problems. Gonadotropin-releasing hormone analogs work wonders. The drugs shut down puberty for as long as they are taken and allow a kid to remain a kid until they reach the appropriate age to start growing up.

"We've seen absolutely no mental problems in any of the precocious-puberty patients, so why should we see any problems in transgender kids? Even so, some medical systems in some countries are nervous and delay treatment until kids are well into their teens, but by that time it is just too late to be effective," Dr. Spack pointed out.

For females seeking to become male, androgens, such as testosterone, are slowly introduced to cause the body to develop masculine features such as facial hair, larger muscles, big hands, tall stature, and a deep voice.[61] For males seeking to become females, estrogens are administered to soften the skin, redistribute body fat to places such as the hips, reduce muscle mass, and develop breasts.[62] The effects are dramatic.

"One of my male-to-female patients is now a beauty queen in the UK," said Dr. Spack, beaming.

It is work that Ovid could only have dreamed of . . . transformation into the opposite sex. If such transformations reduce the number of young people who kill themselves in despair, well, that sounds like magic of the best kind to me.

# IMMORTALITY AND LONGEVITY

## Elixirs of Life, Holy Grails, and Distilled Alchemist Pee

*Look who knows so much. It just so happens your friend here is only "mostly" dead.*

—MIRACLE MAX, *THE PRINCESS BRIDE*

China's first emperor, Qin Shi Huang, ruled from 221 BC until he died in 210 BC. During his eleven years in power, he arranged for the completion of the Great Wall, had his minions build the vast terra-cotta army, and passed dozens of reforms that not only cohesively unified China but set the stage for emperors to rule with stability for centuries. Industrious, organized, creative, and ruthless, he was one of China's most important leaders.[1] He was also just a little bit crazy.

I don't like the idea of dying. Nobody does, but Qin Shi Huang took his fear of death to a whole different level. He knew the Taoist legends* of *hsien*, people who had attained a form of immortality, and consulted the famed alchemists Ch'i Sheng and Hsü Fu so that he could follow in the footsteps of these supernatural souls.

The alchemists told him of three types of *hsien*. The first, or lowest grade, was the *shih chieh hsien*, or "corpse-free immortal." The bodies of these *hsien* simply vanished when immortality was attained.[2] In many cases these *hsien* simply left behind their clothes as they disappeared,

---

* Which the Chinese delightfully call "wild histories."

but sometimes they left behind their bones as well. These immortals did not ascend to heaven but neither did they die. They simply remained, whispering in the ears of their loved ones for eternity. If you think this all sounds a bit like what happens to Obi-Wan Kenobi in *Star Wars*, you are not alone. George Lucas got a lot of his ideas from Asian culture, including the samurai-like helmet worn by Darth Vader.

The second grade was the *ti hsien*, an earthly immortal who appeared to be a living and breathing person but was free to roam the lands without aging for all eternity. The third and highest grade were the *t'ien hsien*, individuals who transformed into celestial beings and lived much like gods in the heavens.[3]

Qin Shi Huang initially aimed for option two. He used his forces to explore a legend of five mystical islands in the Po Sea. These islands rested on the heads of fifteen gargantuan turtles until a great giant came and killed six of them to use their shells for divination magic.[4] As a result of this celestial turtle harvesting, two of the five islands floated away and sank near the North Pole. The three islands that remained, which were named P'eng-lai, Fang-chan, and Ying Chou, were inhabited by animals of purest white and a group of *ti hsien*, who maintained their immortality through the cultivation of a magical herb.

Convinced that the islands were real, Qin Shi Huang sent the alchemist Ch'i Sheng off to track them down. With three thousand people at his command, Ch'i Sheng's expedition was formidable.[5] It never returned.

Some believed the alchemist and his associates found the islands, attained immortality, and decided that returning to China wasn't worth it. Some whispered that they never found the islands and were fearful of being executed if they came back to the emperor empty-handed. Others suggested the expedition became lost at sea and that everyone died. Today, scholars argue that the expedition may never have happened—that it was simply a myth invented by the Chinese to explain how the

Japanese islands came to be inhabited. Regardless, Qin Shi Huang was not deterred. Facing extreme old age,* he turned to local substances.[6]

## TURTLE POWER

At the time, a lot of odd materials were consumed to ward off aging. Some, such as dog urine, insect blood, and black cow dung, were downright disgusting. Others, such as the eggs of cranes, which were thought to grant a thousand years of extra life, made some mythological sense, since cranes were viewed as messengers between gods and mortals.[7] A few, such as broth made from tortoiseshell, were logical.

Many turtles and tortoises live for a long time.[8] Three-toed box turtles are well-known for living longer than sixty-five years; the oldest Blanding's turtle is documented to have lived to at least seventy-seven years; and Lonesome George, the last of the Pinta Island Galápagos tortoises, died in 2012 at the age of at least one hundred.**[9] By modern standards, where the average life span in developed countries often hovers around eighty, such longevity in wild animals is fascinating. But consider what our ancestors would have thought when average life spans were far shorter. People living 10,200 years ago averaged twenty-one years of life, Egyptians in AD 40 averaged twenty-six, and the English at the end of the sixteenth century averaged forty-two.***[10] That turtles and tortoises could easily outlive two generations of people was nothing short of miraculous. It followed that consuming their various bits increased life.

Today we know that eating turtle toes or sucking up broth from a tortoiseshell does not extend life, but this has not stopped labs from try-

---

* Which back then was forty-seven.

** He is thought to have been a fair bit older, since he was already an adult when discovered in 1912.

*** One thing to keep in mind is that once people survived all of the diseases of childhood, they stood a greatly increased chance of living into their forties, fifties, or sixties. More on this later, though.

ing to unravel the mystery of turtle longevity. One theory is that their long lives are related to their aquatic lifestyles.[11]

Many turtles' brains have the extraordinary ability to survive for long periods without oxygen. Turtles can live about three or four days without oxygen in 77°F (25°C) water, but in 37°F (2.7°C) water they can survive without oxygen for months.[12] When turtles do come up for air, they get a big burst of oxygen into their tissues and endure this experience unscathed.

Why is a big burst of oxygen something to worry about? When oxygen enters animal tissue, highly reactive oxygen ions and peroxides are released. These materials are extremely aggressive and damage DNA. This DNA damage, often called oxidative stress, is thought to play a key role in aging. Following their first gulp of oxygen-filled air in months, turtles do not seem to experience a burst of oxidative stress.[13]

Some researchers argue that this is because turtles mop up reactive oxygen in their bodies before it causes DNA damage by having higher levels of antioxidants, which fight oxidative stress.[14] Others argue that turtles are not so much mopping up reactive oxygen as they are resisting its effects through exceptionally effective DNA repair systems.[15] It should come as no surprise that these sorts of mechanisms for protecting DNA drew a lot of attention from governments when fear of nuclear war was at an all-time high during the twentieth century.

In the wake of World War II, the US government grew concerned about protecting its soldiers from the DNA-damaging effects of nuclear radiation. The government backed research initiatives and drug development, which ultimately led to a promising discovery by a team of researchers led by David Grdina, a radiation biologist at the University of Chicago.[16]

Dr. Grdina gave neonate mice a compound known by the ridiculously long chemical name of S-2-(3-aminopropylamino)-ethylphosphorothioic acid. Just thirty minutes later, he and his colleagues exposed some mice to a single dose of radiation that was twenty times greater than annual maximum exposure rates established by most US federal agencies. The team examined the tissues of the mice for evidence of cellular damage and reported that S-2-(3-aminopropylamino)-ethylphosphorothioic acid provided significant protection against radiation.[17]

The drug that Dr. Grdina tested is now known as amifostine, and it seems to work by increasing production of MnSOD, a major antioxidant enzyme found in the tiny powerhouses inside cells called mitochondria. This increase was on the order of ten to eighty times greater than normal and made the cells much more effective at neutralizing reactive chemicals before they could do any harm to DNA.[18]

Today, amifostine is administered to some cancer patients who are intentionally being exposed to radiation to destroy their tumors.[19] The drug allows patients to protect themselves from harmful chemicals just as turtles do. The big question is whether such a drug could provide protection against the daily doses of highly reactive chemicals that cause us to age. Experiments in this area have yet to be run, but they are in the cards.

Bruce Carnes, a gerontologist at the University of Oklahoma, argues that amifostine's proven radioprotective effects make it a strong candidate for protecting against aging and that a short-term pilot study is warranted. Such a study, he points out, would only need two groups of mice to be examined, one receiving amifostine at a dose known from the cancer research to have radioprotective effects and a control group. Animals would enter the study at an age near the end of their reproductive lives; have their blood, feces, urine, and saliva examined at a molecular level regularly; and then, after four months, be sacrificed so their organs could be analyzed. Dr. Carnes proposes that if the amifostine provides a day-to-day protective effect against the highly reactive oxygen that our bodies encounter through normal activity, the organs of the mice taking the amifostine should look healthier and less aged than the organs of the control mice.

We do not know if Qin Shi Huang dined on turtle soup or drank dog urine as he raged against his mortality, but we do know that he ultimately turned his attention upon the earth.

*Tan sha*, or red sand, was a key ingredient in Taoist elixirs of immortality.[20] The idea that consuming such stuff was good for you seems crazy given that red sand is highly toxic mercuric sulfide. However, a closer look at what they were doing to prepare *tan sha* for use in their potions starts to explain why it was viewed with such awe.

*Tan sha* was prepared by roasting, which released what the Taoists called living metal.[21] This living metal was pure mercury, the only metal that can remain liquid at room temperature. For a kid in chemistry class seeing it for the first time, it does look remarkable and somewhat alive as it rolls around on a platter. Yet, pure mercury is not particularly safe.

Mercury readily reacts to the element selenium, and enzymes composed of selenium protect our brains from reactive oxygen produced by normal metabolic activity. Mercury irreversibly shuts down these enzymes, leaves brain cells vulnerable, and sets the stage for extensive brain-cell death. When hatmakers used mercury to treat the animal furs that they made hats from, they regularly inhaled mercury vapors. While many such hatmakers merely became pathologically shy, depressed, and forgetful, some became permanently delirious. Thus the phrase *mad as a hatter* and the creation of the Mad Hatter in Lewis Carroll's *Alice in Wonderland*. Yet, in spite of all the dangers of mercury, it is not lethal to drink in its pure form.

"When little kids ingest mercury from broken thermometers, their parents freak out because they think it is going to kill them," explained Harvard Medical School toxicologist Alan Woolf as I sat with him for a casual tea in August of 2014. "The metal just runs harmlessly through the digestive system. We usually find it in their diapers later on." This flew in the face of everything I knew about the element, so I double-checked the matter with University of Massachusetts Medical School toxicologist Edward Boyer, and he confirmed what Dr. Woolf had said. "Drinking mercury is safe. It is inhaling mercury compounds that is highly toxic."

My meeting with Dr. Woolf left me fascinated and confused. Qin Shi Huang was believed to have died from consuming mercury drawn from the magical red sands of China.[22] I had initially accepted this as fact since, like so many others, I understood that elemental mercury was deadly. However, if Dr. Woolf is correct, then the legendary emperor died of something else. Perhaps another toxic material ended up in his magical elixir? We may never know.

## THE PHILOSOPHER'S STONE

Qin Shi Huang's story is dramatic, but it is just one of many. In ancient India, there were stories of a northern land known as Uttarakurus. The people in this place ate fruit from a tree that allowed them to live for a thousand years free of disease.[23] The Babylonians tell in their legend of King Gilgamesh that, after he witnessed the death of his friend Enkidu, he became obsessed with obtaining immortality. He tried mastering sleep, believing that if he could stay awake for six days and seven nights, he would pass a magical threshold and live forever. When that didn't pan out, he tried to find a plant growing at the bottom of the sea that would grant immortality. Unfortunately, when he got close to the plant, a snake snuck in and ate it before he could pick it.[24] This element of the story may have been added as an explanation for why snakes shed their skins and appear to cheat death by periodically rejuvenating themselves.

The ancient Greek doctor Hippocrates and many of his followers believed that aging was the result of people's losing a vital moisture in their bodies. To combat this, they suggested that people consume materials such as gold, coral, and pearl.[25] Roger Bacon, an English philosopher and alchemist from the thirteenth century, proposed that while consuming small amounts of gold was good, breathing in the breath of youths was even better.

By Bacon's day, people widely knew that inhaling the breath of a seriously ill person could lead to the contraction of that person's illness. With this level of understanding, people reasoned that the reverse was also likely true. Breathing in the breath of healthy people would grant health.[26] Thus began a popular, not to mention mildly disturbing, custom in which old men paid to replenish their youth by inhaling the breaths of virgin girls.*

During the 1600s, consuming gold was still viewed as a pathway to immortality, and the purer the gold, the better. Alchemists toiled long and hard to create the right sort of pure gold, but many believed that this was impossible without the use of a mystical material. While most people today know this material best as the magical stone that Voldemort tried to steal from the dungeons of Hogwarts in *Harry Potter and the Sorcerer's Stone*,**[27] the philosopher's stone was actually an ingredient much sought after by our ancestors.

The stone had many forms. In some texts it is described as a purifying powder that could be applied to gold during heating. In others, it was an actual stone that could be dropped into a solution when gold was being soaked with other chemicals. If the very purest of gold could be produced from the philosopher's stone, it could then be consumed to gain eternal life.

In 1669, the German alchemist Hennig Brandt was working intensely toward his goal of creating the legendary stone in his Hamburg lab.[28] He started by distilling fifty buckets of urine through intense heating. This reduced the urine to a thick syrup and allowed a red oil to come to the surface.[29] He drew this oil off and left the rest to cool, the upper part developing into a black and spongy crust, and the bottom congealing into a thick and salty liquid.*** Brandt discarded the salty

---

* Remarkably, there is no evidence of any clinics offering elderly women the opportunity to pay for the breaths of virgin boys.

** The title of the first edition of the book, published in Britain, was *Harry Potter and the Philosopher's Stone*.

*** You really wish that he had some sort of chemical means of testing that the liquid was salty.

bottom layer, mixed the red oil back into the black, spongy crust, and then boiled the mixture for sixteen hours. White fumes were released, then oil, and finally the solution that remained was immersed in cold water to solidify. The result, Brandt discovered, was a rock that glowed in the dark.[30]

Brandt was intensely secretive about his finding for six years, but eventually, in 1675, he revealed what he had found to the wider alchemical community. Some individuals, such as the alchemist Daniel Kraft, grew quite wealthy showing off their glowing rocks to nobles, but the rocks proved useless for purifying gold.[31] Brandt had not found the life-granting philosopher's stone. Or had he?

Feces and urine are exceptionally effective fertilizers. For nearly five thousand years, the people of China have scattered waste over fields to keep their crops growing. In medieval England, feudal peasants grazed their sheep on the lands of the nobility, but were forced to leave behind their animals' droppings.[32] Excrement is rich in phosphorus, which is an essential ingredient for all life on earth. However, it also spreads illness.

As people came to realize that exposure to feces and urine led to outbreaks of horrible diseases such as cholera, it became critically important to remove waste efficiently.[33] Toilets and sewers sent waste off to oceans and hugely reduced infection rates. These technologies helped people live a lot longer, but they came at a significant cost.

While phosphorus had historically been put back into the soil through excrement, this changed with the invention of sewer systems. Phosphorus started getting flushed out to sea. With waste suddenly less available and human populations getting larger, farmers came under intense pressure to find new sources of phosphorus to keep their crops growing. Concentrated deposits were identified in many localities around the world, and these began to be mined. Use of mined phosphorus shot through the roof.[34] During the population boom that followed World War II, phosphorus production sextupled.

There is no substitute for phosphorus, and the amount that can be mined is limited. The United States, considered the world's largest producer, is expected to have only a few decades left of the stuff. China, Morocco, Jordan, and South Africa, which are the next biggest producers, are similarly constrained. Run out, and worldwide crop production will crash, leaving starvation in its wake.[35]

Mother nature provides some phosphorus return all on her own. Salmon, which feast on phosphorus-consuming creatures in the ocean, spawn in streams and get eaten by bears, which then defecate the phosphorus in meadows and forests. Similarly, numerous fish-eating birds dump phosphorus at inland locations. These pathways are useful but are not enough on their own. While distilling urine to create glowing rocks may not sound like the most appealing of tasks, developing techniques that draw phosphorus out of waste for use on farms will be pivotal in the decades ahead. A philosopher's stone indeed . . .

## THE HOLY GRAIL

The philosopher's stone was not the only magical object that our ancestors thought granted eternal life. The grail used at the Last Supper and that caught the blood of Jesus Christ beneath the crucifix has also featured prominently in immortality mythology.

The concept of a magic goblet is actually associated not with immortality as much as with longevity in nobles, dukes, and kings. People in power make enemies and gain attention in ways that normal folk such as you and I do not. This is why the president of the United States gets carted around in a bulletproof limo, is surrounded by armed guards, and has a team to make sure the pastry chef in the White House is not secretly working for the North Koreans.

Ancient rulers really had trouble on this final point. They could surround themselves with guards and barricade themselves inside castles, but royals struggled to monitor where the ingredients for their food and drink came from. As *Game of Thrones* makes all too clear, getting poisoned was a perpetual risk. This led many royals both to hire food tasters and to seek out tactics for withstanding toxins. One such tactic was to consume elixirs and substances that granted a supernatural protection against poison. Among the most famous individuals who followed this practice was Mithridates VI, king of Pontus, a region that is now in Turkey.[36]

As a young boy, after watching his father fall prey to poison in 120 BC, Mithridates was determined to avoid the same fate. During his reign he ran hundreds of experiments on prisoners, commoners, friends, and even himself.[37] He started mixing ingredients—including many poisons in tiny doses—into a honey paste that was then crafted into a chewable tablet.[38] He consumed one of these tablets a day throughout much of his life and is widely believed to have rendered himself immune to poisons.[39] When he was about to be captured by his Roman enemies in 63 BC, he attempted to kill himself and his daughters through poison. His daughters dropped dead, but his own body shrugged off the toxins.[40]

If this brings to mind the line "I spent the last few years building up an immunity to iocane powder,"[41] then you are not alone. When I first encountered it, the Mithridates tale felt as if it had to be as fictional as *The Princess Bride*, but there is probably some truth here.

Mythologist Adrienne Mayor, who wrote an entire book on Mithridates entitled *The Poison King*, points out that the body can put up with many poisons in small doses. One example she cites is arsenic. After regular exposure to tiny amounts of arsenic over a long period, the liver adapts by producing enzymes that neutralize the toxins. With enough of these enzymes present, a resistance to the poison builds up, and a person develops a moderate tolerance.[42]

The plant *Hypericum perforatum*, better known as Saint-John's-wort, is listed in many of the recipes that historians suggest Mithridates was using to make his magical tablet. This plant has profound effects on the body. One study, led by molecular endocrinologist Linda Moore at the Glaxo Wellcome Research Center in North Carolina in 2000, revealed that Saint-John's-wort induces the expression of an enzyme called CYP3A4, which oxidizes and removes foreign molecules circulating in the blood.[43] Increasing the body's production of CYP3A4 helps the body to better find and destroy poisons before they can do their damage. Yet, it would be better to not drink poison in the first place, and this is where magical grails come into the picture.

For centuries, arsenic trioxide has been a key concern among people in power because it is so easy to use in assassinations. Like the fictional iocane, it is odorless, tasteless, and dissolves instantly in liquids. Just a pea-size amount in a drink will kill an adult in under two hours.[44] For the more creatively inclined assassin, smaller doses of the poison can be dropped into drinks over several months to create the illusion that a person is falling ill from a disease. I doubt that Mithridates encountered arsenic trioxide, since that lethal compound, known by many as the poison of poisons, wasn't created until the eighth century AD in the laboratory of the noted Arabian alchemist Jabir ibn Hayyan.[45] Nevertheless, by the fifteenth century, arsenic trioxide was so widely used to kill royalty that it came to be known as the powder of succession. For those who had dealings with the Medici family who ruled Florence during that period, the poison was wielded with such skill that those who died after dining with them were said to have been "Italianated."[46]

With death by poison preoccupying the European elite, they sought out protective objects. Bezoars, calcareous stones collected from the stomachs of animals such as goats and sheep, were thought to magically protect against poison. They were dropped into drinks or ground up and poured into them to neutralize poison.[47] Unicorn horns were also highly valued for protection. We know this because we find them in the collections of wealthy rulers such as Lorenzo de' Medici and have records of their being sold as an antidote for ten times their weight in gold.[48] The "unicorn horns" were not horns at all* but the tusks of narwhals, close relatives of dolphins that grow one long tooth out the front of their heads. Might there be something real behind beliefs in bezoars and unicorn horns? I didn't think so, but then I encountered some magical beliefs that struck me as so weird that I just had to look closer.

On the island of Malta, locals believed that the stone and fossilized shark teeth from a cave where Saint Paul once took refuge had magical properties against poison.[49] They believed that if fossil teeth or stones from the cave were ground up and poured into a drink, any poison would be neutralized. Similar magic was associated with cups and stirrers made from the stone. Even stranger, islanders thought the fossilized shark teeth would "sweat" if poison was nearby.[50] It struck me as odd. Why would anyone imagine a magical tooth "sweating" to reveal the presence of poison? How could stone from a cave wall or fossil shark teeth have any effect on poison? The cave on Malta is made of calcium carbonate and the shark teeth are made of calcium phosphate.[51] Narwhal tusks and most stirrers are also calcium based. Was some sort of chemical reaction between calcium and arsenic taking place? I pulled out my periodic table and made the necessary notes. If I added calcium carbonate to arsenic, what would I get? Calcium arsenate?

$$CaCO_3 + As = Ca_3(AsO_4)_2$$

It looked wrong to me, and, being a paleontologist rather than a chemist, I realized I was going to need some help from an expert. Fortunately, my friend Dave is a chemistry whiz.

It was 9:30 a.m. for me in the United Kingdom, but having known Dave for twenty-four years, I figured he'd still be up at 1:30 a.m. in Los Angeles. "Yeah, hi, I'm not waking you, right?" Laughter burst through the phone. Why did I even bother asking? Aside from being far better at chemistry than I am, Dave is also far better at staying up into the wee

---

* Sorry to break the news, but unicorns don't exist.

hours of the night. I quickly moved on to my question and realized that I'd written in elemental arsenic rather than arsenic trioxide. Aside from that, Dave pointed out some other matters that I'd forgotten.

"You're gonna need some oxygen in there to begin with too, and the reaction is going to release some carbon dioxide. You cool with that?" asked Dave.

Wine and other beverages certainly had oxygen dissolved in them, and if carbon dioxide was released, it would just bubble out of solution. "Sure," I replied.

"Then, yeah, you could make that work, I think." A few minutes of mutterings and technical debates with his wife, Elodie, also a chemist, led to a solution:

$$3CaCO_3 + As_2O_3 + O_2 = Ca_3(AsO_4)_2 + 3CO_2$$

"If you had a cup made of calcium carbonate and you poured a liquid poisoned with arsenic into it, the arsenic would get drawn out of solution and you'd get a layer of calcium arsenate forming on the surface of the cup," explained Dave.

"But would that really stop you from being poisoned?" asked Elodie in the background.

"Seriously? If I knew I had to drink arsenic, I'd rather have one of those calcium carbonate cups, thank you very much!" retorted Dave. Even so, their debate made me wonder. Would there truly be a protective effect? Curious, I again turned to Dr. Woolf.

"The big question is how bioavailable arsenic is after it bonds to a chemical like calcium," Dr. Woolf said. "It would be a strong bond for sure, but I honestly don't know if it would be strong enough to keep the arsenic from causing trouble. It would be easy enough to test, though."

Four months later I found myself on the Boston metro with a box of arsenic trioxide in my hands. It was disturbing to know I had enough poison in my hands to kill everyone else in my train car. Fortunately, the journey didn't take too long, and I had the poison of poisons safely behind lock and key an hour later.

Working in close collaboration with toxicologist Edward Boyer at University of Massachusetts Medical School and environmental engineers Harry Hemond and Irene Hu at MIT, I arranged to test whether our ancestors might have been onto something with their unicorn horns, bezoars, fossil shark teeth, and grails. We all knew that calcium carbonate was used by modern water-treatment plants to scrub arsenic out of city water supplies, but we weren't sure if merely drinking out of a stone cup could provide protection. For four days Dr. Hemond, Irene, and I slaved away inside the lab poisoning different sorts of wines, swishing the toxic solutions with various stones and fossilized teeth, and mixing them with powdered calcium carbonate before running them through analysis machines.* Charlie, one of my old friends from university, even went to the effort of carving a grail out of limestone for a final test.

The whole experience was as entertaining as it was educational. First of all, we learned that arsenic trioxide does not dissolve instantly in water or wine. It barely dissolves at all. Every time Irene dropped a pea-size amount of the poison into a flask, it stubbornly floated at the top. Stirring didn't help much either. Even if we hadn't labeled our flasks carefully, we would still have known which flasks were poisoned due to the white powder on the surface. The second thing we learned was that neither calcium carbonate powder nor the presence of limestone reduces the presence of the arsenic in solution.

The only thing that did make a difference to arsenic levels was iron oxide. Irene and I had read that iron oxide, better known as rust, could precipitate arsenic out of solution. We added some rusty water to a few of our experiments and did see lower levels of the poison in solution afterward. That was cool but it didn't support the mythology. As for

---

* A full video of our antics is available at www.amatteroffactandfiction.com.

fossilized shark teeth sweating in the presence of poison, we tested that too. We kept staring at teeth dangled over toxic wine but never saw a single droplet of moisture form on them.

It is unclear what, if anything, we did wrong. Possibly something is chemically present in bezoars, narwhal tusks, and the limestone of Malta that interacts differently with arsenic from the calcium carbonate and fossilized teeth that we worked with. Possibly the arsenic we had was different from the arsenic that was used to murder people hundreds of years ago. We used arsenic trioxide, but maybe arsenic pentoxide would have interacted differently? Perhaps there was another poison altogether? Possibly the chemistry of the drink that the arsenic was in could be a factor. While we tried to find wines that were created using historical techniques, they might not have been similar to the wines that were being poisoned so long ago.

So was there a holy grail that could grant immortality? Not really. The nobles at risk of being poisoned might have come to believe that certain cups, stirrers, and stones could grant them longer lives, but I suspect anything more than that had to have been fiction.

## METHUSELAH MICE

Biblical stories abound of people living to seemingly impossible old age. After leaving the Garden of Eden, Adam is said to have lived for 930 years.[52] Noah lived for 950 years, and his grandfather Methuselah lived for 960 years.[53] These tales raise the tantalizing question of just what is possible when it comes to lengthening human life.

In 1900, the average life span in the United States was roughly forty-five years, but it is now closer to seventy-eight years.[54] In one hundred years, a mere blip in the history of the world, thirty-three years have been added to the average US life span. That's a whopping 73 percent increase! Many people would like to believe we will be able to do this incredible feat again.

Some argue that this rise in life span has come almost entirely from defeating contagious diseases, such as diphtheria, measles, and tuberculosis, that once frequently killed children. They suggest that with these causes of mortality now gone, we will not see life spans climb again in the same way.

This is a fair argument. In 1900, nearly 15 percent of all babies died before their first birthdays. Today that figure is closer to 0.06 percent. Mathematically, when calculating the average life span of a population, losing an infant dramatically reduces the overall average. Thus, a pool of lost lives was pulling average life spans down. As soon as that pool was eradicated with sanitation, antibiotics, and improved nutrition, life-span averages leapt upward.[55] Whether there will be another leap in longevity depends upon what sorts of treatments modern medicine yields.

In 1998, James Thomson at the University of Wisconsin and John Gearhart at Johns Hopkins University revealed that they had taken germ cells found inside embryos and which produce sperm and eggs and transformed them into stem cells.[56] This was a big deal, because stem cells have the potential to become just about any cell type in the body. They will one day make it possible to grow organs in a laboratory for transplant patients.[57] Stem cells could eventually be injected into worn-out parts of the body to rejuvenate aging tissues.[58] Such techniques are being seriously considered for treating the brains of people suffering from diseases such as Parkinson's and Alzheimer's.[59] It was all exciting stuff, but there was, alas, an ethical problem.

Initially, the best way to provide a person with a steady supply of rejuvenating stem cells was to clone the person and collect stem cells from the clone in its embryonic state. As you can imagine, a lot of people had problems with this. Was it right to grow an embryonic clone solely to harvest its tissues? The question conjured up creepy images of humans being grown like crops. However, the question suddenly stopped being discussed in 2006 when biologist Shinya Yamanaka at Kyoto University found a way to transform mature cells found in adults into stem cells.[60] His discovery sent shock waves through the world and won him the 2012 Noble Prize in medicine.

Would stem cells help to increase life spans? I turned to an expert on aging. "Yes, but not by much," said Jay Olshansky, professor of public health at the University of Illinois at Chicago. The reason, he argued, is that problems associated with aging cannot be treated in the way that diseases normally are. "If you have someone with a failing liver at forty, growing them a new one with stem cells is going to grant many more years of life. In contrast, growing one for someone at eighty is less useful because, while the new liver will do well, the rest of the body is not going to last long." The trick, he said, is directly interfering with aging rather than treating the conditions it causes.

Amifostine (discussed earlier) has potential, he explained. As for a drug called rapamycin, which was shown in a 2009 issue of *Nature* to increase female-mouse life span by 14 percent and male-mouse life span by 9 percent, "the evidence so far is very promising," said Dr. Olshansky.[61] Rapamycin meddles with a biochemical pathway that senses nutrients. When an animal's food supply is reduced, this pathway slows down and aging decelerates to a crawl.[62] Why this happens is not known for certain, but evolutionary biologists speculate it is about buying time during periods of food scarcity.

Consider two mice, one in a forest where food is plentiful and one in a forest suffering from drought. For the mouse where food is readily available, breeding, which demands a lot of calories,* can happen quickly and easily. For the mouse living where food is scarce, breeding is out of the question. But that mouse is not entirely out of luck. While the mouse in the forest of plenty will have a normal life span of about

---

* Just think about how many tubs of Ben & Jerry's a pregnant woman can get through and you'll understand what I'm talking about.

eight hundred days, the hungry mouse will live 50 percent longer. This extended life results from nutrient-sensing pathways slowing down and the body going into full preservation mode so the mouse can increase her chances of living until food becomes available again and pregnancy can follow.

Rapamycin lies to the body, telling it that food is scarce and pushing it into a partial preservation mode that increases life. I know, it sounds amazing. So why aren't we all taking it? The reason is safety. Aside from messing with nutrient-sensing pathways, rapamycin is also an immunosuppressant. Taking it could make people more vulnerable to disease. "[Although] there is a lot of potential here, we really need clinical trials," said Dr. Olshansky.

Beyond drugs, genetic manipulation may also increase longevity. "We can already engineer mice that have these nutrient-sensing pathways altered so that they automatically live 50 percent longer than normal mice—that's 1,200 days. If we also put them on extremely limited diets, simulating an environment where food is scarce, their lives reach 1,600 days. Our Methuselah mouse is 1,825 days old. That's like 160 years if you translate things across to people," explained gerontologist Valter Longo at the University of Southern California in Los Angeles.

Genetically manipulating these pathways leads to near-perfect immunity to cancer and diabetes. If you think I am doing a bit of wild speculation here, you're wrong. Some human populations in the world have naturally occurring mutations in their nutrient-sensing pathways that have remarkable effects. "It's amazing, really. There's a population of three hundred people with this mutation in Ecuador, and in all our years of studying them, none have ever developed diabetes, and only one has ever died of cancer," said Dr. Longo.

Nobody is sure if the mutation actually helps people live longer, though. "It's hard to tell. The people with the mutation in Ecuador drink and smoke lots. I even asked one of them, 'Don't you worry about your health?' And he was, like, 'Nah, I'm immortal.' More than twenty percent of them die in accidents . . . like getting hit by cars. It's crazy. If they lived clean and went on food-restricted diets, they could be breaking all kinds of longevity records," said Dr. Longo.

How those who do not get killed in accidents actually die is a mystery. It has proved difficult for doctors to find anything that could have been fatal. "They just drop dead for seemingly no reason. It's the same with our genetically engineered mice. When we open them up, we often just can't figure out what the cause of death was," explained Dr. Longo.

Initial studies on monkeys suggest that controlling diets can increase longevity in primates too. In one study, published in *Science* in 2010, a team found that reducing the normal diets of rhesus monkeys by 30 percent for twenty years led these monkeys to experience 50 percent fewer incidences of cancer and heart disease than a control group of rhesus monkeys that were left on normal levels of monkey chow.[63] Moreover, while 42 percent of the control monkeys developed diabetes during their lives, none of the monkeys on the restricted diet did so.

Should we start restricting our diets by 30 percent to live longer too? Maybe, but some dangers here have not yet been explored. For example, cutting back on nutrients could potentially make people more vulnerable to infections by weakening their immune systems. While disorders associated with old age may be fended off by restricted diets, this wouldn't matter much if calorie restriction increased the odds of a person's catching influenza and dying at thirty.

But there may be another way forward. In 2008, a team led by gerontologist Luigi Fontana at the Washington University School of Medicine in St. Louis published an article in the journal *Aging Cell*.[64] They discovered that while nutrient-sensing pathways in rodents were reduced in activity by 30 to 40 percent when food was restricted, this was not so in people unless protein in the diet was also restricted.[65] Dr. Longo has followed up on this research and published an article in *Cell*

*Metabolism* in 2014 stating that protein restriction in a human population led people to experience a 40 percent reduction in mortality risk from all causes and a 75 percent reduction in cancer death compared to a population on a high-protein diet.[66]

"Through a combination of drugs and diets we are on the verge of seeing a lot more healthy one-hundred-year-olds in the years ahead," said Dr. Longo.

Dr. Olshansky is more cautious about what the future will bring: "People have constantly asked, 'When is the next big breakthrough going to happen?'—and the answer has been 'Ten years from now' for, like, forty years. I think we really have the potential to expand healthy years of life so that many more people make it to eighty in good shape and then rapidly decline just before they go, but increasing average life span beyond that, well, I think we're only going to be crawling forward at this point."

Dr. Longo disagrees: "We are crawling only because we are doing such a poor job of preventing people from aging themselves both early and late in life. If we can generate mice that consistently live forty percent longer with half of them remaining highly functional and disease-free until they die, we will eventually be able to get humans to live longer too."

Regardless of whether we will see average life spans leap by a further twenty years or just see people get healthier up until age eighty, both options would have looked like magic to China's first emperor.

# SUPERNATURAL SKIES

Sunstones, Plagues, the Power of Planets,
and Summoning Storms

*A philosopher once asked, "Are we human because we gaze at the stars,*
*or do we gaze at them because we are human?" Pointless, really . . .*
*"Do the stars gaze back?" Now that's a question.*

—NARRATOR, *STARDUST*

There is nothing magical about waking up at 4:00 a.m. There is
nothing magical about driving at that hour either, particularly if you
find yourself on the twisting lanes of London in the dark. Even so, my
friend Charlie was visiting us from San Francisco for the New Year, and
while Thalia had to work, he and I were setting out to find magic. We
were not disappointed.

"What was that?" asked Charlie as the first flash lit up the world
around us. I hadn't noticed. Perhaps it was just the headlights of another
car? Then the sky answered that question with a roar. A few large rain-
drops splashed onto the windshield. No big deal; I'd driven in rain
before. Yeah, right.

Thirty minutes later, our windshield wipers were at full throttle.
With the exception of occasional flashes of lightning, neither of us
could see a damn thing in the wall of water.

After an hour, the storm vanished as quickly as it had appeared. The
skies cleared and began their transformation from black to the dull gray
of near dawn. We arrived at a parking lot surrounded by grassy fields

and sheep for as far as the eye could see. I showed my press credentials to a guard, and in moments we were walking up a muddy path.

As I caught my first glimpse of Stonehenge, I felt let down. Even as I walked up closer, I was unimpressed. So they were rocks stacked upon one another. What was the big deal? As I studied the site more carefully, I realized it was distance that was diminishing the power of the place. I glanced to my side. The handful of other visitors were all sticking to the boardwalk, yet I had arranged to walk inside the stone circle. I showed the guard my paperwork. He looked a bit startled at first but then muttered something into his radio. Moments later, he got a garbled response and told me I could go in. So I did.

When it was first constructed in 3000 BC, Stonehenge was just a circular ditch. Only around 2500 BC did the huge stones start arriving. The largest, known as sarsen stones, weigh over thirty-five tons and were brought from the Marlborough Downs in North Wiltshire, 19 miles away. The smaller stones, known collectively as the bluestones, are from the Preseli Hills in Wales, over 150 miles west of Stonehenge.[1]

There is a lot of debate over how the stones were brought to the site in an era without wheels and erected into their present formation without elephants or cranes.[2] Some argue that they were dragged, some suggest wooden sleds were used, and some hint that they were floated along the River Avon for part of their journey.* There is even more debate over why our ancestors went to all the effort.

Geoffrey of Monmouth, a cleric during the 1100s who was more of an author of British legends than a historian, suggested the Welsh bluestones had healing powers when water was poured over them. Archaeologists Timothy Darvill at Bournemouth University and Geoffrey Wainwright at the Society of Antiquaries of London found ancient

---

* One disastrous attempt by archaeologists to reconstruct the bluestone journey led to a team of volunteers dragging a three-ton stone seventeen miles through Wales, only to watch it sink off the coast just after it was shoved aboard a raft.

art carved on rocks at springs in Wales that supported this idea. Given that these springs were located where the bluestones were collected, the two suggested that the stone circle was used to conduct ancient curative rituals.[3] Gerald Hawkins, an astronomer at Boston University, argued that the whole site was an astronomical calendar built to predict lunar eclipses.[4] Mike Parker Pearson, a lecturer at the University College London Institute of Archaeology, suggested that different circles may have been created for different groups, with the wood circle that once stood having been built for the living and the stone circles having been built for the dead.[5] He also argued that the stone circles built from local stones and the stone circles built from Welsh stones may have represented different ancestors, and that the placement of the two circles together was an ancient attempt at merging ancestral lines. Perhaps the most eccentric argument is the one based upon sound.

Paul Devereux, an expert in acoustics and archaeology at the Royal College of Art in London, analyzed the bluestones and found they emitted sounds that were not dissimilar to those made by bells.[6] When Devereux and his colleagues looked at the stones more closely, they found evidence that they had a long history of being physically struck. Their theory is that these rocks were once used as a giant instrument to play tunes for ancient audiences.

I don't mean to dismiss the academic battles, but as I stood there staring at the size of the rocks right in front of me, none of it seemed to matter. As with the storm that we had just driven through, something about standing in the center of the circle made me feel insignificant. The magic of the place really came alive then. As I marveled at my own tiny spot in the world, the sky to the east caught fire and a ray of light launched its way through a massive arch. A vermilion sun started to rise, beautifully bordered by the stones. All I could do was gasp in awe.

Given their visibility and ability to evoke wonder, it is hardly surprising that the sun and the moon frequently appear in our mythology.

In Norse legends, the sun goddess, Sol, traveled through the sky pursued by the hungry wolf Sköll. At night, the moon god was similarly chased by Sköll's lupine brother Hati. On some days Sköll got worryingly close to eating Sol. Indeed, the Vikings speculated that one day Sol would be caught, and when that happened, a terrible battle in which gods would die, known as Ragnarök, would follow.[7] Scientifically speaking, the Vikings were more right than they could ever have realized. One day, a long time from now, the sun will die, and the effect on our planet is not going to be good.

The Egyptians were not much different from the Vikings. They believed that their sun god, Ra, led a boat crewed by gods across the sky every day, and that every night Ra returned to the eastern point where the sun always rose by traveling through the underworld. This was a dangerous journey, because Apep, a vile serpent god, frequently tried to ambush Ra.[8] Like the Vikings, the Egyptians told stories of moments when one of Apep's ambushes proved particularly effective.

Were the moments when Sköll and Apep were getting the upper hand eclipses? Were the Vikings and Egyptians seeing such things? I was curious, so I started working with eclipse calculators and was surprised by just how many eclipses our ancestors would have seen. Over the city of Giza between 1500 and 1000 BC alone there were four annular eclipses, in which the moon gets in front of the sun to create an eerie ring-of-fire effect, and hundreds of partial eclipses, in which the sun would have looked as if a bite had been taken out of it.[9] Similarly, the Vikings would have seen an annular eclipse over Oslo on September 9, 945; an annular eclipse over Stockholm on May 17, 961; and a total eclipse over Stockholm on July 20, 966. Like the Egyptians, the Vikings also would have seen hundreds of partial eclipses.[10] How did they react? We can't know for certain, but my guess is that they looked on in fear and told stories to their children and their children's children about the terrible moment when the world almost came to an end. Regardless of whether eclipses were actually behind these myths, evidence is emerging that the celestial bodies really do hold significant sway over our lives.

## LUNACY

In the spring of 2013 I had just come back to Europe from a two-week teaching stint in the United States and caught a train to Brussels, where I was looking forward to a long, relaxing weekend with Thalia and some of her extended family. The next morning I woke up jet-lagged and blearily walked into a wall. As I stumbled downstairs, a few of my littlest relatives beckoned me to play with LEGOs with them, and I gave it my best effort.

Their mother pointed out that my mood probably wasn't so much about jet lag or travel exhaustion as it was about my relationship to the planets. Apparently, something astrological that she had been reading said it was going to be a bad day for an Aries such as me. So the stars were aligned against me too. Fantastic.*

Months later, I had all but forgotten about my little encounter with astrology when a fascinating article in *Current Biology* landed on my desk. The paper suggested that at least one celestial body, the moon, did hold sway over our lives.[11]

Like most good science, this exploration, led by chronobiologist Christian Cajochen at the University of Basel in Switzerland, started one evening in a bar. Dr. Cajochen and the other researchers were having a beer in 2011, under the light of a full moon, when a question came to them: Can the phases of the moon have some measurable effect on humanity? They

---

* Later that day I went to the local swimming pool with two of my nieces and had my swimming suit rejected by the lifeguards. Yeah, you read that right: *rejected*. They were just ordinary swimming shorts, but this was a Speedo-only pool. Seriously? I was given a choice: either don't go swimming, or borrow the lifeguard's ill-fitting Speedo. Unhappily, I went with option two.

knew that in a bedroom with windows, the increased light from a full moon would disrupt sleep, and that the darkness from no moon would improve it. Beyond this, they found themselves wondering, if light is removed from the equation, might the phase of the moon still have an influence?

In 1995, in the *Journal of Biological Rhythms*, physiologists Martin Wikelski and Michaela Hau at the Max Planck Institute for Behavioral Physiology in Germany found evidence of what they called a circalunar clock in reptiles.[12] The work was conducted with the Galápagos marine iguana, which grazes on algae in areas often covered with ocean water that become exposed at low tide. Dr. Wikelski and Dr. Hau suspected that the iguanas somehow knew when low tides were coming. They found that the iguanas often started walking to intertidal feeding grounds from their inland resting areas hours before low tide took place. Even iguanas that had fasted for several days still arrived at feeding areas right at low tide as they broke their fasts. Aware that these fasting iguanas could be making use of social or environmental cues, the researchers ran a series of experiments with the animals in enclosures where they could neither see the sky, get in ocean water, or interact with other iguanas that had recently visited the sea. Astonishingly, the activity levels of the iguanas remained tightly bound to nearby tidal activity that they could not possibly see, hear, or smell. The researchers concluded that the iguanas had some innate sense of where the moon was.

Why the moon? Because the moon plays a big role in controlling tides. When the moon is right over our heads, it creates high tides right where we are and on the exact opposite side of the planet too. In contrast, it creates low tides in areas that are ninety degrees from us on the planet.

Since the earth rotates 180 degrees in twelve hours while the moon rotates only 6 degrees during this same amount of time, the moon should be over our heads roughly every twelve and a half hours. With this in mind, Dr. Wikelski and Dr. Hau speculated that the iguanas somehow sensed when the moon was either six hours away from being overhead or six hours past being overhead and thus creating a low tide that would allow them to graze on algae. How the reptiles were doing this was a mystery, but the team theorized that the iguanas have the lunar equivalent of the circadian clock that helps us to instinctively remain in tune to the day and night cycles that rule our lives.

Curious if humans might unknowingly have a similar lunar sense, Dr. Cajochen and his colleagues started pondering how to design an experiment to explore this. They were lucky. Between June 17, 2000, and December 2, 2003, they had brought seventeen healthy young volunteers and sixteen healthy older volunteers into the laboratory to study how the body regulated sleep. That initial study had never aimed to explore the influence of different moon phases on sleep regulation. As a result, the idea of the moon having an effect on sleep was never mentioned to the participants, the technicians, or even the senior researchers guiding the study. It was effectively better than a double-blind experiment; it was an entirely blind one.

The experiment consisted of two nights in a sleep laboratory followed either by forty hours of sleep deprivation or forty hours of relaxation made up of awake time and naps. Both of these conditions were then followed by eight hours of sleep.

When the researchers went back and looked at the different moon phases and considered the initial experiment's findings in relation to this information, they discovered something surprising. Around the time of the full moon, non-REM sleep, the deep kind, in which we tend to dream, decreased by 30 percent on average. It took participants five minutes longer to fall asleep during the full moon as well, even though they were in enclosed rooms. Sleep duration during the night of the full moon also fell by twenty minutes on average. Moreover, these changes were associated with a decrease in the level of sleep quality that participants believed they were getting and were tightly connected to lowered levels of melatonin, the hormone in our bodies that helps to keep control of our sleep/wake cycles.

Dr. Cajochen and his colleagues reported that these changes in sleep pattern affect how we live.[13] They can shape how we feel and treat one another. Since the results suggested that these sleep disruptions happened in most people, they could be responsible for large behavior patterns seen in humanity. Might studying increased overall irritability in society during periods of the full moon reveal something?*

To date, Dr. Cajochen's experiment has been replicated three times by other researchers. Twice the results have mirrored what he found.[14] Once, they did not.[15] Such findings are encouraging, but more work needs to be done. If the results are real, we're left with the huge question of why the moon has this effect.

Dr. Cajochen and his associates accept that some moonlight, seen by study participants before they arrived at the lab in the evening, might have had an influence. Could just seeing the phase of the moon before sleeping in an entirely enclosed space have such a significant effect? That seems doubtful. This led the researchers to wonder if the effect could be gravitational.

The sun, like the moon, has an effect on the tides. Its effect is nowhere near as powerful because it is much farther away. Even so, the sun's effect is significant enough for us to notice. Why am I talking about the effect of the sun when Dr. Cajochen and his colleagues made their findings about the stages of the moon? It's because the stages of the moon are related to the sun.

During the full and new moons, when the earth, moon, and sun are in alignment, the tides become particularly high and particularly low. In contrast, when the sun and moon are separated by ninety degrees and the moon appears quartered, the tides are weaker.

These forces are probably not having some sort of gravitational effect

---

* It makes me wonder, if we were to look at major outbreaks of violence over the past five decades, would we see more episodes of rioting and military aggression at the time of the full moon than at other times of the lunar cycle?

on us. While the moon and the sun do influence oceanic tides, they do not produce tides in most small lakes and definitely do not influence the tiny pools of water found in our own bodies.

But it is possible that, like the iguanas, we have some sort of lunar clock ticking in our bodies in relation to the positions of the moon and the sun. This may allow us to, subconsciously, know when a full moon is approaching. As I read through the *Current Biology* findings, the paleontologist in me was screaming the evolutionary question "Why would an ability to sense the full moon ever be selected for by evolution in people?"

Again, much can be learned from Dr. Wikelski and Dr. Hau's work with the iguanas. In their Galápagos field study, which spanned three years, they found that marine iguanas with the most accurate lunar clocks were the ones that were the most likely to survive during seasons when food was limited. While the researchers could not know why this was, they theorized that the animals with the most accurate lunar clocks reached feeding spots first and gorged themselves on the most algae.* Having an accurate lunar clock and following it granted a distinct advantage. Could the same sort of thing have once been true for people? Yes.

An archaeological study led by Curtis Marean at Arizona State University and published in *Nature* in 2007 revealed the discovery of 160,000-year-old shells and stone tools at the South African coastal site of Pinnacle Point.[16] Located roughly halfway between Cape Town and Port Elizabeth, Pinnacle Point is one of the earliest locations where humans were regularly turning to the ocean for food. Indeed, our ancestors left behind quite a mess for us to find. Why they were suddenly seeking out shellfish is a matter of debate. One of the most convincing arguments is that, as the climate became cooler and drier at that time, food on land became more scarce and the sea needed to be exploited for nutrients.

Archaeologists have discovered heaps of shells belonging to the spe-

---

* Just like the saying "The early bird gets the worm," only with algae-eating iguanas.

cies *Perna perna*, better known as the brown mussel, amid piles of ancient garbage. This suggests the people were eating, and likely depending upon, shellfish dwelling along the coast. If this was so for thousands of years, and it looks as if it was, then that window of time is long enough for evolution to hold sway over humanity. What would this evolution have looked like? Well, I'd argue for the iguana effect, where people born with a sensitivity to the lunar cycle gained first pick of mussels at the lowest tides of the month. In effect, I am suggesting that the early human catches the mussel while the late human with a poor lunar clock goes hungry. Such a theory is far from proven, but it seems plausible.

Has all of the lunar work led me to become a devout astrologer? No. I still have a hard time accepting that the positions of celestial bodies when we are born lead us to encounter good or bad events. What I am coming to believe is that a kernel of truth lies behind all of this. In the future, I suspect we are going to discover that the positions of the sun and the moon have many more effects on us than we currently realize. Of course, these celestial bodies do far more than just subtly influence our lives; they have also played a key role in guiding us across vast distances.

## PRESCIENT PEBBLE

The Norse Hrafns saga tells of a meeting between King Olaf the Holy and the son of a farmer named Sigurd.[17] Sigurd boasts to the king that he always knows where the sun, moon, and stars are, even if he is stuck in a bank of fog or stranded in the midst of a blizzard. Fascinated by this, the king asks for a demonstration, since it is snowing heavily. Sigurd agrees, goes outside, and points in a specific direction, claiming that the sun is there. The king takes a magic stone from his robes, holds it to the sky, and confirms the true direction of the sun by the way the stone gleams. Just a myth, right? I wasn't so sure when I first perused this tale.

The Vikings were adept seafarers. They started traveling from Norway to Iceland in AD 860. They then went on to Greenland in AD 980, and to Newfoundland around AD 1000.[18] This alone is amazing, considering that they were making their journeys in wooden ships that were only around 54 feet (16.5 meters) in length, but what is truly staggering is that they managed all of this long before the magnetic compass was used in Europe.[19] Without a compass the trips could potentially have been achieved by keeping an eye on the sun and using a device attached to the deck called a horizon board, which had the bearings for sunrise and sunset marked on it with a set of wooden pegs.[20] However, for such a navigational tactic to work, the skies over the northern seas would had to have remained clear for most of the voyage for the sun to be properly tracked, and that isn't likely. This has left a lot of scholars baffled and led some to speculate that King Olaf's stone was not quite as fictional as it sounds.

A few minerals interact with light in strange ways that can help identify the location of the sun when the weather is not behaving. One of the most notable is a clear crystal made of calcite called Iceland spar. When an ordinary beam of light hits the crystal, the beam gets split in two.[21] The resulting pair of beams head off in different directions and make anything seen through the clear crystal appear as if it has been duplicated. For example, if I take a piece of Iceland spar and stick a square of black tape to one side of it, I would see two images of the tape if I looked at it through the crystal.

The spar reacts in an unusual manner with the light generated by the sun at dawn and dusk, making it useful for navigation. At these times of day, the light waves that make up the sun's rays tend to be linearly polarized, like the light from a computer screen, after having passed through lots of the earth's atmosphere. The duplicate images that are seen through the crystal do not appear to be of equal intensity unless the optical axis of the crystal, oriented toward the sun's zenith, is rotated forty-five degrees from the direction of polarization of the sky at the zenith. So the argument goes that the Vikings were studying the double image seen in a crystal to help them spot the location of the sun on cloudy days.[22]

For decades, the argument had no evidence to support it; then a chunk of Iceland spar was discovered among navigational instruments on an Elizabethan ship that sank in 1592.[23] This wasn't a Viking ship, but it existed almost two hundred years before scientists even understood the concept of using a crystal to determine direction at dawn and dusk. A team of physicists led by Guy Ropars at the University of Rennes in France got ahold of the Iceland spar from the wreck for analysis and found it to be a cloudy mess after having been sitting at the bottom of the sea for centuries. Nevertheless, Dr. Ropars and his team were able to source a near-perfect replica of the specimen and test out its optical properties. Sure enough, they reported in *Proceedings of the Royal Society* in 2011, with the crystal they could determine the location of the sun within just a few degrees even when it was far out of sight below the horizon.[24]

Dr. Ropars's paper created quite a stir. Some naysayers countered that the Vikings made their journeys during the summer, when the sun would rarely have approached the horizon, rendering the spar ineffective.[25] Dr. Ropars disagrees with this argument, countering that "it is precisely in summer when twilights can last for several hours." Another argument is that the crystal technique would have been useless in heavy fog or during snowstorms. Again Dr. Ropars points out that this misses a key point: "Even a brief and small blue patch of sky at the sun's zenith would be enough during the day to use the sunstone."

Unsure of whom to believe, and being an enthusiastic sailor myself, I decided to give my friend Sarah at the British Geological Survey a call. As expected, when I told her what I was working on, she was jubilant and certain she could help me source a crystal. A month later we were aboard a sailboat off the coast of Turkey with a chunk of Iceland spar in our hands. The sun was just starting to dip below the horizon, and the sky was clear. It was the perfect time to test the sunstone. We had stuck a small square of black tape to one side of the crystal and were excitedly gazing through it in the twilight. We spotted the duplicate image of the tape in moments, and as we expected, the images were not of equal strength. We aimed the long axis of the Iceland spar at the sun's zenith and looked again. The images kind of equalized, but not really. We tried again, this time aiming the sunstone in different directions to see if we were missing something. We messed with the stone until it was dark, and then again during each of the following days. Sadly, we remained unconvinced. Toward the end of our little field experiment we postulated about what we were doing wrong. Was our crystal too small? Was it somehow impure? Was the latitude off the coast of Turkey causing a problem? Were we aiming the stone incorrectly?

Not one to give up easily, Sarah obtained a huge, pure chunk of Iceland spar a month later and tested the thing way up north near Arthur's Seat in Scotland with her partner, Gareth. The evening was clear and the sun hovering just at the horizon. Again the conditions were perfect for testing the spar, and again the mineral did not prove convincing. By the end of their testing, Gareth was certain he could use the stone to detect the sun's direction, saying that the stone needed to be tilted slightly to see the duplicate-image tonal changes. Sarah couldn't see what he was talking about.

Given our experiences, I suspect practice is needed to wield a sunstone effectively. Dr. Ropars and his colleagues seem to have put in the time and figured it out. With more work, I expect that Sarah and I will too. Indeed, after we reported our failure to Dr. Ropars, he confessed himself as confused as we were. Perhaps a visit to his lab in France is in order to get the hang of using the stone. Even so, I don't think the myth of King Olaf having a stone that "gleamed" in the direction of the sun during a snowstorm is true, since the stone does not "gleam" at all, but I can believe that the Vikings considered minerals such as Iceland spar supernatural.

The celestial bodies, however, were not the only things that our ancestors perceived to be associated with magic from above.

## LET MY PEOPLE GO

The book of Exodus in the Bible describes a series of plagues that were sent by God to encourage the Egyptians to let their Hebrew slaves go free. In the ninth plague, darkness descended upon Egypt for three days. The Bible says, "Then the Lord said to Moses, 'Stretch out your hand toward heaven so that there may be a darkness over the land of Egypt, a darkness that can be felt.' So Moses stretched out his hand toward heaven, and there was dense darkness in all the land of Egypt for three days."[26]

The plague of darkness has sometimes been attributed to an eclipse, but in recent years scholars have started coming up with more interesting arguments. While there is huge debate over the origins of the Exodus narrative and whether there even were Hebrew slaves trying to escape Egypt, most scholars agree that Exodus itself is a mix of stories that were passed down orally for well over five hundred years before it was ever written down.[27] Orally passed along for hundreds of years? That sounded like a recipe for disaster when I started studying the magical elements in the tale, but, as with a lot of magic, when I looked closer, things were not what they initially seemed.

With books and movies, we have a physical object that stores the beliefs and wishes of the people who were alive when the book or movie was initially made. Even if the story gets re-created at a later date with new themes, the original material is still there for us to reference. As an example, consider the 1962 Spider-Man comic and the 2012 film *The Amazing Spider-Man*.[28] In the original story, Peter Parker becomes Spider-Man after being bitten by a radioactive spider, but in the more recent film he becomes a superhero after being bitten by a genetically modified spider. The story has changed over fifty years to reflect the anxieties of the day, with the radiation fears of the 1960s being swapped out for modern genetic engineering concerns. Moreover, we can see the story's evolution by comparing the two forms of the Spider-Man narrative. We do not have this luxury with Exodus, since its earliest forms were all orally transmitted. Nevertheless, we must not think that oral storytelling drastically changes narratives in the same way that a game of telephone mutates messages whispered from one person to another around a campfire.

Contrary to popular belief, oral storytelling preserves tales rather well. We know this because of work done between 1934 and 1950 by two clever Harvard classicists, Milman Parry and Albert Lord. They were curious about *The Odyssey* and *The Iliad*, which, like Exodus, had been oral stories long before they were ever written down. Had people memorized these epics? That seemed doubtful to the scholars, given that the works were over twenty-one thousand lines long, but they weren't sure. To find out, Dr. Parry and Dr. Lord turned their attention to one of the few places in the world where audiences still listened to oral epics: Yugoslavia.[29]

They found bards there who would sing tales to people gathered in cafés. Crucially, some of the tales told by one talented bard they met, named Avdo Mededovic, were just as long as *The Iliad* and *The Odyssey*. These tales were rarely written down, which was just as well, since Avdo and most other singers were illiterate.[30] Even so, the bards did not have their epics memorized in the classic sense. Dr. Parry and Dr. Lord discovered the bards were stringing stock phrases, narrative fragments, and formulas together to create a story. When the scholars recorded two performances of the same tale by the same bard, they found that the details of each performance were unique, but that the overall narrative was nearly identical. The same was true of the same tale being told by two different bards. Peripheral details sometimes changed, but the story did not. Narratives were locked in their form. This was true even when stories recorded by Dr. Parry in 1934 were compared to stories recorded by Dr. Lord in 1950 after Dr. Parry had died. The passing of the years had little effect.[31]

All of this discussion of epics in Yugoslavia is relevant to Exodus because it can help us identify which material in the story is likely to have changed. So, with that in mind, did some natural disasters inspire the ten plagues?

Serious scientific explanations for the Exodus plagues began in the 1940s. Archaeologists John B. Garstang and his son John Garstang proposed that the plagues were a result of a volcanic eruption in the Rift Valley of East Africa.[32] They argued that the eruption dumped toxins into the lakes that fed the Nile. These toxins, they suggested, killed the fish and flushed sickened frogs inland, creating plague number one, the Nile's turning to blood, and plague number two, a frog infestation.[33] They further argued that insects swarmed the dead fish and frogs, bringing about plague number three, biting insects, and plague number five, disease. As for the later plagues of hail and darkness, the team proposed that a second volcanic eruption at a different location blasted vast ash clouds over Egypt, creating these.[34]

These arguments ran into geological trouble because the volcanoes in the Rift Valley tend to erupt basalt,[35] which is best known as the runny molten material erupted on the Hawaiian Islands. Such eruptions can sometimes release toxins into water sources, but this is rare. Crucially, the delta is almost four thousand miles away from the lakes that feed the Nile near the Rift Valley, and the Nile itself has a very large number of tributaries that would have diluted any potential toxins with clean water before they got to Egypt.

In 1957 Dutch physician Greta Hort proposed that powerful summer floods from alpine lakes in Ethiopia brought red-colored protists, known as flagellates, into the Nile.[36] She suggested that these flagellates made the river look bloody, sucked up the oxygen dissolved in the water, and killed all the fish as a result. The protists drove frogs inland,

while the high flood levels brought clouds of mosquitoes, thus causing plagues two and three. Under such miserable conditions, infections
became rampant, causing the fifth plague of disease. Then her argument
turned a bit fuzzy. A random hailstorm wreaked havoc, locust swarms
flew in from Arabia, and a freak dust storm brought darkness.

Critics argued that powerful summer floods from Ethiopia were
common and asked why such a regular event would be considered
supernatural. These reasonable criticisms didn't stop more ideas from
being proposed, though.

In 1964, Greek seismologist Angelos Galanopoulos proposed that
the eruption of Thera, a volcano on the Greek island of Santorini,
released ash clouds that led to the events that caused the ten plagues.[37]

According to geological evidence, the Thera eruption ranks as a
whopping 7 on the 8-point Volcanic Explosivity Index.[38] What exactly
does a ranking of 7 mean? The technical description found in the index
is "mega-colossal." Such eruptions send more than a hundred cubic
kilometers of ejecta into the air and create plumes larger than forty kilometers in size.[39] They are, in short, utterly devastating.

The Thera eruption is the second-largest eruption to have taken place
during the past four thousand years. The largest, of Tambora in Indonesia
in 1815, launched so much crap into the air that it effectively blocked the
northern hemisphere's summer that year. Such events are minor compared to category 8 eruptions, wherein volcanoes the size of small countries cause what geologists describe as "apocalyptic" effects around the
globe. Fortunately, our recent ancestors did not see any of these.

Over the years, Dr. Galanopoulos's argument has gained traction.
Most recently, Barbara Sivertsen, managing editor of the *Journal of
Geology* and an expert on how oral storytelling shapes narratives over
time, put her mind to teasing out the science behind the plagues.[40]

Sivertsen notes that at the moment of the Thera eruption, the Egyptians would have heard a distant rumbling and possibly even felt a shock

wave. The volcano would have triggered tsunamis up to 39.4 feet (12 meters) high. These waves would have contaminated freshwater sources. This would have created a disaster by killing freshwater fish and making a lot of water undrinkable. Would it have made the water look like blood?

Thera ejected a lot of ash during the months before it finally blew its top, and this ejection laced much of the surrounding ocean with both iron oxide and sulfuric acid. Iron rusts when it enters water and tints it red. Further, sulfuric acid partially dissolves iron in water and allows a number of toxic algae to readily consume it. Sivertsen argues that red, algae-filled ocean water got forced up the Nile when the tsunamis hit. She suggests that plague two, the frog infestation, followed swiftly thereafter, since frogs would flee rivers and ponds if their watery habitats were so completely corrupted.

As for the biting insects and flies found in the third and fourth plagues, Sivertsen posits that insects might not have been involved at all. In Exodus, Moses's brother Aaron strikes the ground with his staff, and the dust that gets stirred up transforms into gnats.[41] As the ash clouds from Thera reached the Nile, the lightest bits of ash would have started to fall down like dust. Volcanic ash is acidic, and when it lands on the skin, it irritates it much like insect bites. Might this description of "dust irritating the skin *like* biting insects" have changed over time to become "skin irritated by biting insects"? As stories pass orally through the generations, people try to make sense of elements in them that seem contradictory.[42] Thus, if an original version sung by a bard contained dust falling from the sky, this detail would have changed over the years as other bards learned the tale, since dust usually comes from the ground.

In the wake of Thera, many animals in the fields would have developed respiratory problems, become ill, and died from ash getting into the mucous coatings in their lungs. Alternatively, pathologist Regina Schoental at the Royal Veterinary College in London proposed in 1984 that contaminated water, along with thousands of dead frogs and swarming insects, would have presented a pathway for diseases to spread rapidly among livestock, and that this killed many of them off.[43] Both seem reasonable explanations for the fifth plague of animal disease. With the sixth plague, in which Moses tosses soot from a furnace into the air and it becomes blisters on the Egyptians' skin, this is likely a continuation of the acid-burning effects of the falling ash.[44]

As for the violent hail, thunder, and fire in the sky that came with the seventh plague, might this have been the result of the electrically charged ash clouds colliding with naturally occurring storms? Such a collision could have caused icy balls of acidic ash, lightning, and a blackout that lasted for days. Indeed, this could be where the ninth plague of darkness comes from.

As for the locust plague that arrived between the catastrophic hail and the darkness, these insects are described in Exodus as having come about because "the Lord brought an east wind upon all the land all that day and all that night; when morning came, the east wind had brought the locusts."[45]

Sivertsen argues that winds from the southern edge of the Mediterranean could easily bring locusts into Egypt.[46]

The final plague, the death of the firstborn Egyptians, is perhaps the most difficult to explain in any sort of naturalistic manner. Epidemiologists John Marr and Curtis Malloy proposed in the journal *Caduceus* in 1996 that the combination of previous plague events might have tainted food supplies with toxic fungi.[47] In Egypt, firstborn sons had priority for food. Thus, if Drs. Marr and Malloy's theory is correct, it may have seemed as if the firstborn were being selectively killed off by some supernatural force.

These explorations of Exodus are fun, but may not be accurate. Ice cores drilled in Greenland in 2005 reveal ash from the Thera eruption that dates to 1642 BC, give or take five years.[48] This date clashes with others found in Exodus. For example, Exodus says that the cities of Pithom and Ramesses were completed by the Hebrew slaves. Egyptian records unearthed by archaeologists make it clear that these cities were built during the reign of the pharaohs Seti I and Ramesses II between 1291 BC and 1212 BC.[49] Based upon these dates, the ten plagues would have come almost four hundred years *before* these cities were built. That's a problem. One possible explanation for these conflicts is that the ice core data is flawed and that the 1642 BC ash is actually from an Alaskan volcanic eruption rather than a Greek one.[50] Alternatively, the conflicting information may have arisen because bards used city names that existed at the time they were telling tales to describe places that once had different names.[51] Even so, I think we gain clarity if we avoid trying to view Exodus as a history text and take Dr. Parry's and Dr. Lord's findings into consideration.

The plagues are the quintessential narrative fragment. With a sequence running from one to ten and a set progression of horrible events, they were the perfect sort of thing for an oral teller of tales to have learned after a real catastrophe, like the Thera eruption, and then dropped into oral stories during the centuries that followed. The plague narrative would have changed little. For this reason I suspect the plague tales that ultimately got written down were inspired by real events but were not necessarily connected to a Hebrew exodus from Egypt.

As for that magical moment when Moses drew upon the power of God to part the Red Sea, there is a lot of debate over what sort of real event might have inspired this story. One theory, proposed by University of Indiana geologist Dorothy Vitaliano, suggests that the parting took place at what are now known as the Bitter Lakes and Lake Timsah. Dr. Vitaliano, who invented the word *geomythology* to describe the practice of searching for science in myths, suggested that during ancient times a canal connected these Mediterranean lakes.[52] When a steady eastern gale blows in, a ridge in the shallow water between the Great Bitter Lake and the Little Bitter Lake gets exposed.[53] Egyptologist Frank Yurco even recounted that he had personally seen the wind blow water across the marshy land around these lakes and thought it highly plausible that a strong wind could expose dry land in a flood area.[54] Sivertsen argues that this is the same sort of weather system that would have brought locusts into the Nile region.[55] I suspect the parting of the Red Sea is another narrative fragment. However, I do not think we are ever going to know whether it was originally connected to the chaos following the Thera eruption or whether it was another isolated event that got picked up by the bards and ultimately woven into biblical mythology.

## CALLING LIGHTNING

All scientific explanations aside, Exodus is quite clear about where the power behind the plagues is coming from. While Moses sets some of the supernatural events in motion, it is always God who makes them happen. Now we are seeing people perform the unbelievable all on their own.

For millennia, we have feared lightning. Part of this fear came from a lack of understanding. Part came from viewing its effects: forest fires and, on occasion, electrocution. This all changed in 1752 when one of America's founding fathers had the bright idea of flying a kite with a key attached to it up into a thunderstorm.

Benjamin Franklin's work initiated a golden age of electrical understanding.* Yet, we still lack the ability to harness the power of lightning. Dr. Emmett Brown summed it up rather nicely in the film *Back to the Future* when he said, "I'm sorry, but the only power source capable of generating 1.21 gigawatts of electricity is a bolt of lightning. . . . Unfortunately, you never know when or where it's ever gonna strike."

---

* There is a lot of controversy over whether Benjamin Franklin merely came up with the idea of drawing down lightning with a kite or actually tested it out himself. Whether he did or did not run this experiment, his interest in exploring it scientifically heralded a change in our perception of lightning from terrifying thing of the gods to something that could be tamed and, perhaps, put to use one day.

While the science associated with the time travel in that film was not science at all, Dr. Brown's attempts to harness a lightning strike hinted at both our inability to do such a thing and our interest in ultimately doing so.[56] That was thirty years ago, though, and while I still don't have my hoverboard, I'm pleased to report that science is taking steps toward making Dr. Brown's lightning work a reality.

Storms are thought to become electrically charged when water, ice, and wind break apart charged particles in the clouds. This process, we think, leads some areas of storms to become loaded with negatively charged particles, and some areas to become loaded with positively charged ones. These charges build up as the water, ice, and wind continue to violently swirl, then this enormous electrical potential eventually drives light, negatively charged particles called electrons to leap toward clusters of positively charged particles, creating lightning. Most of these leaps are from one cloud to another, but sometimes the nearest reservoir of oppositely charged particles is on the ground.

The research that is guiding Dr. Brown's work toward reality is being led by Jérôme Kasparian at the University of Geneva in Switzerland. He and his team are learning how to call down lightning bolts from passing storms.[57] Dr. Kasparian's experiments are much like Benjamin Franklin's, only cooler.

Franklin, if he actually did run his legendary experiment, would have attracted lightning by flying his kite into a thunderstorm. An important part of his work depended upon rain. The kite string had to be wet, since water is an excellent electrical conductor. Dr. Kasparian is effectively doing the same thing but with a laser.

Lasers are capable of breaking apart particles in the air much as swirling rain and ice do. Yet, unlike storms, which break apart particles in a large area, lasers do this in a narrow one. Dr. Kasparian and his colleagues speculated that they might be able to tear away negatively charged electrons near storms. They figured that these newly released electrons would function a lot like Benjamin Franklin's wet string and draw charged particles in thunderstorm clouds down to earth.

To test out their ideas, the researchers went to the ridge of South Baldy Peak in New Mexico and used a Teramobile femtosecond-terawatt laser to fire 1.2-inch (3-centimeter) beams into passing thunderstorms. The laser shot out a beam for 150 femtoseconds—that's one-tenth of a millionth of a millionth of a second—every 100 milliseconds. More important, it was releasing a few terawatts* of power with each of these blasts. This proved enough to break free a lot of negatively charged particles from molecules in the air several hundred meters above and draw positively charged particles in the storm down along them toward the laser.

The result, published in *Optics Express* in 2008, was corona discharges, multicolored flickering lights well known to seafarers as the eerie and ominous Saint Elmo's fire, which frequently appears on the masts and sails of boats just as they are about to be struck by lightning.[58]

Thus, the laser was not doing quite enough to bring a bolt to the ground, but Dr. Kasparian and his colleagues were close. They are tinkering with their laser so they can make it more powerful and truly call lightning down from the clouds in a manner that would make even Thor jealous.

They are putting in all the effort not for show or for powering a flux capacitor but to develop lightning shielding for aircraft. For the moment, the only way to guide a lightning bolt to an object is to fire a cabled rocket into a storm. If the rocket gets struck, the electricity travels down the cable and into whatever testing object it is attached to. If

---

* While we can all come together and agree that a few terawatts is bigger than 1.21 gigawatts, let's all be honest with ourselves and accept that we don't really understand what "a few terawatts" of power is. Technically speaking, it is a few trillion watts, but again, that doesn't say much. What does say a lot is Dr. Kasparian's description: "Matt, just think of all the power delivered by all the power plants in the world for a fraction of a second."

it doesn't, the only option is to fire off more rockets and hope that one eventually gets struck. It is cumbersome stuff, not to mention expensive. Lasers would be better.

## SUMMONING STORMS

Native Americans have a long history of performing dances to encourage their gods into behaving in specific ways. There were healing dances, dances to drive away evil spirits, and dances to grant people protection, but none is more famous than the rain dance.

Rain dances were elaborate affairs that made use of feathers, which were representative of wind, and blue and green stones, such as turquoise and serpentine, which stood for water. Such props would be incorporated into costumes and jewelry and waved about during the dance in the hope that the gods would notice and send over the needed weather.

I have a lot of respect for the power of dance. Like music, it can inspire and raise morale. Doing these sorts of things, as I mentioned earlier with regard to ancient healing practices, can have biological effects, but I'm doubtful that dancing can summon rain.

Skepticism aside, weather summoning is no longer restricted to the stuff of deity-oriented rituals. Today, frozen carbon dioxide pellets, better known as dry ice, and silver iodide can be shot into clouds or dropped into them from aircraft to draw out water vapor.[59] This can cause rain to fall when it otherwise might not or make light rainstorms into heavier ones.[60] A few places have put these methods to use. Notably, China used them to force incoming storms to rain out much of their water before they reached Beijing during the days before the opening ceremonies of the 2008 Olympics.[61] Even so, chucking dry ice and silver iodide into the sky regularly is more expensive than just building pipes and canals to move water where it needs to go.[62] For this reason, most countries don't use these tactics all that often.

One problem that canals and pipes cannot solve is the havoc wreaked by powerful storms. Shooting stuff into hurricanes and typhoons is useless, as they are too strong, but more creative tactics are being developed.[63]

Hurricanes form in tropical areas a little ways away from the equator where warm sea surfaces, a moist atmosphere, and a convergence of winds on the ocean surface feed air upward. This upward motion drives the formation of a system that can reinforce itself in a positive feedback loop as more warm and wet air is drawn toward the low-pressure center. This process creates the powerful winds and intense rainfall that are commonly found in hurricanes.

Alan Gadian, a climatologist at the University of Leeds in Britain, knew that the surface of the ocean was just not warm enough in many parts of the world for hurricanes to form. Part of this was due to location—the Arctic and Antarctica do not get enough direct sunlight, for instance—but another reason was cloud cover. Most of the time, nearly 30 percent of the ocean has low-lying clouds over it that prevent the water below from heating up much. With this in mind, Dr. Gadian theorized that if some of this cloud cover could be made just a little thicker, the result would be weaker hurricanes.[64]

Using silver iodide and frozen carbon dioxide to do this would be prohibitively expensive, so Dr. Gadian and his colleagues speculated that sea spray could be put to use instead. Droplets of brine, thrown high up, are known to have an effect on clouds similar to that of the more expensive chemicals. Thus, Dr. Gadian proposed building unmanned water-spouting vessels powered by the wind and controlled by satellite.

Unlike calling lightning down with a laser in the middle of nowhere, where messing up only means cooking some expensive hardware,* meddling with ocean water on a grand scale in a region where hurricanes are spawned could have large-scale catastrophic consequences if something

---

* And possibly frying a few PhD students.

went wrong. For this reason, Dr. Gadian sent his wind-powered ships sailing out into the processors of his lab's computers to test his ideas.

Based upon his models, Dr. Gadian reported in the journal *Atmospheric Science Letters* in 2012 that if two thousand of his wind-powered brine-spouting ships were sent to make clouds thicker and more reflective of light in hurricane-forming regions, temperatures in those areas would drop by a whopping 7°F (about 4°C).[65] That is enough—more than enough—to deal a mighty blow against the formation of these storms. Furthermore, while building and deploying two thousand ships sounds as if it might be an expensive endeavor, he estimates it would cost about $3 billion, with maintenance charges of around 20 percent of that per annum. This has to be considered in light of the damage that severe storms leave in their wake. Hurricane Katrina delivered $81 billion worth of property damage and killed 1,833 people.[66] And that was just one storm. Throw in the costs of Hurricane Rita and Hurricane Wilma, and the cost of Dr. Gadian's fleet starts to look like peanuts.

However, just as some cancer-killing drugs can also kill healthy cells in the body, subduing hurricanes using a vast fleet of brine-spraying ships could cause side effects. Dr. Gadian and his colleagues admit that if all hurricane-spawning regions of the world were aggressively cooled, dramatic reductions in rainfall could occur in such places as Southeast Asia and the Amazon Basin. A loss of more than one millimeter of rain a day would be likely in these areas, and the team argues that this would be unacceptable.[67] Indeed, the costs saved by weakening hurricanes would be offset by damage caused by droughts and fires in places that would lose their typically heavy rain.

To work around this problem, the researchers tested a more moderate tactic. Instead of deploying ships in all hurricane-spawning regions, they placed them in their computer simulator in a few important ones. Specifically, they put them to work in patches of ocean off the coasts of California, Peru, and Angola, a mere 5 percent of the earth's oceans. This resulted in a surface-water temperature decrease of just 0.2°F (about 0.1°C). That is not much. It isn't even enough to weaken the hurricanes and typhoons that the world already experiences. However, when Dr. Gadian and his colleagues looked at how this tiny temperature change would shape climate patterns in the decades ahead, it revealed something intriguing.

Models built by climate researchers make it clear that as global carbon dioxide levels rise, ocean temperatures are also going to rise, and this is going to make hurricanes more powerful than they already are. Deploying brine-spraying ships in the strategically selected locations would nullify this effect, holding the strength of storms in check and preventing them from becoming worse.

Selling this idea to the countries of the world, which is going to be essential if Dr. Gadian's ships are to ever set sail in real water, will be hard. Given geopolitical complications, even a small trial run may never happen. That would be sad, but it does not change the fact that what our ancestors once believed could only be accomplished with dancing, feathers, and turquoise is now reality. We are capable of creating and controlling weather, even storms as powerful as hurricanes. Just reflect upon that for a moment before we move on to the ways in which plants and animals have woven their way into the magic of our mythology.

# ANIMALS AND PLANTS
# AS OMENS, GUIDES, AND GODS

Odin's Wolves and Ravens, Death Cats,
Talking Forests, and Birds That Predict the Weather

*Eywa has heard you!*

—NEYTIRI, *AVATAR*

It was a crisp August evening as I opened the door of the rental car and stepped out onto the dried grass. The sun was dipping toward the horizon, and the bison were grazing in the valley. With the exception of the half circle of cars at the base of the trail and the road we'd driven in on, there wasn't a human structure for as far as the eye could see. Wilderness untamed. It was a thing of beauty.

"Matt, what is here?" inquired my eleven-year-old niece as she threw on her jacket.

"Wolves, if we're lucky," I whispered back.

Much about Yellowstone is majestic—the geysers, the vast undisturbed forests, the wild, winding rivers—but for me the highlight has always been the carnivores. This is less true for Thalia. Scared to death of grizzly bears and not keen on wolves either, she pulled our nieces near and shuffled down the trail behind me.

As we emerged from the shrubs on the hillside, we saw a small cluster of wolf watchers huddled together with their expensive viewing paraphernalia. They were clearly excited about something.

"We saw the alpha pass through just a few minutes ago, and the ravens following her are still there on the hillside," explained one elderly, bearded gentleman. We wolf-watched with them until nightfall, but didn't spot anything other than a few ravens in the distance and more bison. Even so, the man's comment stuck with me as I drove back to our campground that night.

Wolves and ravens . . . something mythical connected the two but I couldn't remember what it was. The next evening, as we listened to a ranger's campfire talk on wolf-and-raven ecology, I remembered.

According to Viking mythology, Odin was the master of two ravens in his celestial realm of Asgard. Named Hugin, meaning "thoughtful," and Munin, meaning "mindful," these birds gathered information in the mortal world of Midgard and carried messages to Odin's followers.[1] Odin was so tightly connected to these ravens that his people often called him Rafnagud, the raven god.[2] Odin also had two wolf companions, Gere, the greedy, and Freke, the voracious. Due to this link to their god, the Vikings viewed wolf sightings as a good omen indicating that Odin was near.[3]

I'd always thought it strange that the Vikings viewed wolves in such a positive light. Wolves eat a lot of the large mammals that our ancestors hunted. They were our competitors, not our allies. None of this mythology had ever made any sense. Then, as the ranger explained recent research findings in Yellowstone, I started to understand.

After having been absent from Yellowstone for nearly seventy years, wolves were reintroduced to the park in 1995 to try to repair an ecosystem that had fallen out of balance. With wolves gone, elk populations had boomed. Their huge numbers wiped out a lot of the plant life, and many smaller herbivores that depended upon these plants, such as beavers, became scarce in the park. Upon the return of the wolves, the ecology of the park started to return to the richer form that it had long ago. Among the most interesting phenomena that rangers spotted was a bond that formed between the wolves and the ravens of Yellowstone. The ravens seemed to be following the wolves around. To an extent this made sense.

Ravens are predominantly scavengers. They need to wait for a predator to make a kill before they can swoop in and take a quick nibble. Even so, it was unclear if they were specifically tracking wolves over other predators.

Curious if this was the case, a team led by biologist Daniel Stahler at the University of Vermont and the Yellowstone Center for Resources closely monitored ravens between 1997 and 2000 to determine whether they were following both wolves and coyotes, the other predatory canids in the park. They reported in the journal *Animal Behaviour* in 2002 that the ravens were selectively following wolves.[4] Why they were doing this is a matter of some speculation, but Dr. Stahler suspects that because wolf kills are considerably larger than those made by coyotes, the birds have more food to feed on. The ravens seem to have the ability to differentiate between the animals and selectively follow only wolves.

Yet there was more. When Dr. Stahler and his colleagues placed twenty-five animal carcasses at random locations within wolf territory for ravens to discover, only nine were found by the birds within an hour. In contrast, every single kill made by wolves during the study period was discovered by ravens within less than a minute. No doubt about it, the ravens were depending upon the wolves to find food.

Based upon the 2002 study alone, the wolf-raven relationship appears somewhat parasitic. The wolves do all the hard work and the ravens just steal meat when they can get it. However, it is likely the relationship is more symbiotic. Dr. Stahler and his colleagues point out that, during their many years in Yellowstone, they have on several occasions observed ravens locating and harassing injured elk to help nearby wolves spot the weakened animals and more readily make a kill. The ravens appear to be earning their keep by serving up vulnerable prey to their wolf compatriots.

In another study, Dr. Stahler found playful interactions between wolves and ravens. He spotted the black birds mischievously pulling on wolf tails and mock chasing their pups around den sites.[5] It was incredible and perplexing. As I discussed this with Dr. Stahler, he and I both had a hard time coming up with an evolutionary explanation for why the wolves would tolerate such nonsense from a bird species that often steals meat from their hard-earned kills unless the ravens are providing some sort of valuable service to them. "We can't prove that the wolves are giving more than they are taking because we can't create a controlled environment for an experiment like that in the wilds of Yellowstone, but I have no doubt that the wolves have learned that it is worthwhile to watch the ravens closely," explains Dr. Stahler.

His comment got me thinking. The wolf-raven relationship is not new. Such ecological interactions tend to develop over thousands of years, and if people saw this relationship, it makes sense that they would bond the two animals together in their mythology. Remarkably, this ancient mythology lives on to this very day. While scientists were right to criticize the ludicrous physics found in *Thor: The Dark World*, I was pleased to see a reasonable nod to both zoology and history when Sir Anthony Hopkins as Odin sent his ravens off with a missive from his tower in Asgard. Even so, this still leaves us with the question of why the Vikings viewed these animals as the servants of their beloved god, and why wolves in particular were seen as good omens. I think a section from the *Poetic Edda*, a collection of ancient Norse poetry preserved in the thirteenth-century *Codex Regius* manuscript, provides some clues:

*Hugin and Munin*
*Fly each day*
*Over the spacious earth*
*I fear for Hugin*
*That he come not back*
*Yet more anxious am I for Munin.*

*Gere and Freke*
*Feed the war-faring*

*Triumphant father of hosts;*
*For 'tis with wine only*
*That Odin in arms renowned*
*Is nourished forever.*[6]

Translated from Viking poetic gibberish, this means Odin's ravens fly everywhere. There is concern that they might vanish from sight. The war-faring and triumphant Odin feeds the wolves meat because he is well nourished by just wine.

After reading through Dr. Stahler's papers, this poem left me with suspicions. Why was Odin giving away all his meat to his wolves? Was some of this meant for his followers? Were the Vikings watching ravens from a distance, tracking them across woodlands to determine where packs of wolves were going and then using this information to find large animals for the tribal warriors to hunt before the wolves claimed them? This interpretation alone helps to explain why there is anxiety over losing sight of the ravens. It also pairs well with the rich culture of competition that permeates Viking mythology. Viking heroes and gods vied with one another over just about everything, and it makes sense that Odin would want his followers to contend with his wolves for meat in the wild. However, it is possible that the connection was even tighter.

We know that some tribes in the world today, such as the Hadza of Africa, engage in power scavenging, whereby people drive off predators from a fresh kill to steal the meat.[7] If the Viking hunters followed ravens to recent wolf kills and then power scavenged, the wolves would quite literally be providing Odin's warriors with food. Could this have happened? I was curious, so I ran the question past Dr. Stahler.

"That's a really interesting idea. You know, when we are doing our work in the field, one of the main ways we determine if the wolves have made a kill is by watching the raven activity. It is almost immediate.

I can't really see people scavenging wolf kills that have been out for a while—human stomachs just can't handle that—but I can definitely imagine our ancestors following ravens and opportunistically kicking wolves off carcasses to get at fresh meat," mused Dr. Stahler.

During a dinner at Harvard in October 2014, I ran the same question past Richard Wrangham, a primatologist and expert on the history of human feeding activities. He too was warm to the idea and advised that while power scavenging in Africa offers people only a small window of time to steal a kill before the meat is no longer suitable for human consumption, in northern latitudes the window would be considerably larger given the colder temperatures. When I spoke to Harvard paleoanthropologist Daniel Lieberman on the matter, he added, "We don't find African tribes following ravens to predator kills, but we do see them following vultures and then driving the predators away." Seeing as this was the sort of thing that could be tested, I applied for some grants to determine if I could find wolf kills in the wild using raven sightings alone.

Thus it was that I found myself leaping over icy streams and dodging bison in the wilds of Yellowstone where Dr. Stahler was conducting his research. Unlike Dr. Stahler, I didn't have any aircraft or radio telemetry devices. I just had my eyes and those of Charlie, who agreed to help me out on another one of my madcap missions.

We found our first raven within an hour, and it led us on a wild-goose chase to a Dumpster near some stables in the north of the national park. It was a less-than-exhilarating experience, but one that was quickly forgotten when we spotted a second raven in a meadow in the northeastern corner of the park. As we watched the bird, it launched into the air and flew south. It was rough keeping up with it, but we managed to do so and were blown away by where it led us.

The raven met with several others near the shoulder of a snowy mountain partially covered in pines, and after circling for a few minutes the birds all promptly started to descend toward the edge of a cliff. We couldn't quite see where the ravens were landing, so we headed farther east and were stunned to discover a large group of wolf watchers with all of their binoculars trained on the cliff that the ravens had led us to. Three wolves were in the midst of an attack on a bull elk.

The ravens were eagerly flapping around in the nearby trees like macabre cheerleaders as the elk swiped its rack of antlers at the wolves. He was putting up a fierce fight, and even succeeded in striking one of the wolves hard in the face; but it was a fight that was not going well for the elk. He was bleeding, cornered against the cliff, and outnumbered. Five hours later, the wolves took him down.

It took three full days for all of the flesh to be ripped away from the corpse, and during that time the raven activity made the kill visible to the naked eye from almost a mile away. While the rangers would not allow us to rush in and test whether we could use spears to steal a few pounds of elk meat from the wolves, Dr. Stahler advised that it would not be hard. "Over the years, when I accidentally encountered wolves close up, they have always run away from me, even when feeding," he explained.

Rather remarkably, the elk was not the only big meat source that we traced using ravens. While visiting the Old Faithful region in the west of the park we also found a dead bison on an icy riverbank. Whether the bison had been killed by wolves or died of natural causes was a mystery. Wolves were certainly nearby. Indeed, we spotted two in a meadow a mile away from the corpse just after leaving the site. Regardless of the cause of death, we counted more than fifteen ravens creating chaos as they picked at the flesh, and we were certain that we could have driven them off to grab a chunk of bison meat for dinner. In short, the entire journey was an astonishing and eye-opening adventure.*

The Vikings were not alone in their beliefs. The Old Testament describes the raven as having been sent by God to feed the prophet

---

* Which is available for viewing as a video on www.amatteroffactandfiction.com.

Elijah when he was nearing starvation.[8] A similar tale is found in later material, such as Saint Jerome's *Vitae patrum*.[9] In that tale a raven brought food to the hermit Saint Paul as he lived out his long life in a cave, and when Saint Anthony the Abbot came to visit him, ravens brought the two of them food.[10] True, the food that the birds brought is described as "a loaf of bread and dates" as opposed to freshly slain elk, but the ancient concept of the raven being a provider of food remains.

The concept of the raven as a divine guide is still alive today. I am not sure whether George R. R. Martin is aware of it, but in his Song of Ice and Fire saga, known on television as *Game of Thrones*, he is carrying on Norse mythology when he leads Brandon Stark to his destiny with a raven. What is particularly interesting is that Bran always sees and follows the raven in his dreams while he is in the form of a wolf. That might seem backward, given that ravens typically follow wolves to kills, but according to Dr. Stahler it is not. Wolves are less able to find prey during winters when snowdrifts are deep and visibility is poor, and this leads them to look to ravens for guidance.[11] Under such harsh conditions, ravens search the landscape for weak, injured, and sickly animals that would make easy prey for their wolf allies. Once they find such animals, they create a commotion much like the one we saw that helps the wolves to close in and make a kill. Thus, amid the deep snows far north of Winterfell, it is entirely appropriate that a raven is guiding Bran in wolf form.

Dr. Stahler's report of this behavior left me wondering whether ravens may have once done the same for the Vikings and their ancestors. When they saw human hunters tracking animals in the wild, did they work out that drawing their attention to easy prey meant that they would be able to swoop in for a few bites of fresh meat, as they do with wolves today? We will probably never know, but it seems plausible.

Plenty of myths go the other way, with the raven filling a mythological role as thief, agent of doom, and occasional consort of witches. This too makes sense given the bird's niche. Ravens are opportunists, and while they can certainly lead the way to wolf kills, they will also readily eat human food if it is not well protected. They'll even pluck out the eyeballs of fallen soldiers on fields of battle. Our myths about animals are shaped by our experiences. Learn to follow ravens to food and they are the agents of beloved deities. Foolishly leave food out for them to steal and they become nothing short of fiends.

## PANDORAN REALITY

It isn't just ancient myths that were used to come to grips with astonishing things in the natural world. Modern mythology does this too. As I mentioned earlier, *X-Men*, *The Incredible Hulk*, *Spider-Man*, and *The Fantastic Four* all emerged at a time when we were just starting to comprehend the dangers and benefits of atomic reactions. These fictions effectively provided a place for people to explore a topic that they did not much understand and were rightly anxious about, but modern mythology does not only respond to subjects of apprehension—it sometimes latches on to wondrous bits of pure science.

As a science journalist, I encounter in my day-to-day work a lot of academic papers with titles that leave my head spinning. In 2005, one paper in the journal *Plant Cell Physiology* threw me for a loop with the title "Volatile C6-aldehydes and allo-ocimene activate defense genes and induce resistance against *Botrytis cinerea* in *Arabidopsis thaliana*." Confusing as the title was, I understood enough of the paper to realize that the researchers were reporting on a chemical mechanism that individual plants of a single species, in this case *Arabidopsis thaliana*, used to communicate with one another.[12] It seemed interesting enough, so I ran the idea past my editor, Geoff, along with a few others. Apparently it failed to impress, because I instead ended up getting assigned to write up something on allergies and fruit that week. The paper and the findings got lost in my waste bin of rejected story proposals, and then, a few years later, I stumbled upon something similar in the journal *Oecologia*.

Again the title "Costs and Benefits of Induced Resistance in a Clonal Plant Network" didn't exactly roll off the tongue, but it further explored the idea that plants were communicating with one another and coordinating defensive action against animals that came to munch on their leaves.[13] I thought it was interesting, so I proposed it to Geoff. No luck. Six years passed, and in the early summer of 2013 a high school intern working with me named Benjamin passed along a paper from *Ecology Letters* that took the earlier findings and pushed them to an entirely new level.

The work, led by botanist David Johnson at the University of Aberdeen, was following up on a 2010 discovery by a team of scientists at South China Agricultural University that I had entirely missed.[14] The Chinese team reported in the journal *PloS One* that when a tomato plant was infected with a disease known as leaf blight, other tomato plants nearby started activating genes that would help prevent them from catching the blight, even if the uninfected tomatoes were encased in a protective barrier.[15] This led the team to deduce that some sort of danger signal was being sent by the diseased plant to the others through the ground.[16] They suspected that symbiotic fungi living in the soil, whose hyphae (which are a bit like fungal roots) were providing the tomatoes with minerals in return for nutrients, might be helping to transmit this signal, but they could not prove it.[17]

Dr. Johnson knew from his own past lab work that when broad-bean plants were attacked by aphids, the beans engaged in chemical warfare by releasing irritating compounds that made life unpleasant for the insects. However, he wasn't sure if broad beans under attack informed nearby bean plants of the danger and induced them, in a tomato-like fashion, to also start spewing out noxious chemicals. He set up an experiment to explore this and determine if symbiotic fungi were playing a part.

Eight mini-environments of four broad-bean plants each were grown by Dr. Johnson and his colleagues for four months in the lab. One plant in each of these was encircled by a mesh with holes that were half a micron in size, large enough for water and dissolved chemicals to pass through, but far too tiny for plant roots and fungal hyphae to breach. One plant in each environment was encircled with a forty-micron mesh, which was large enough to allow the plant to make contact with fungal hyphae but still too small for it to make contact with the roots of other bean plants. The remaining two plants were left alone so that one could function as a control, having full root and fungal contact, and one could be infested with aphids.

A little over a month before the end of the experiment, Dr. Johnson encased the plants in special bags that allowed oxygen, carbon dioxide, and water vapor to enter and exit the plant space freely but blocked all other chemical compounds from moving in or out. Then, four days before the experiment was set to finish, one plant in each micro-environment was infested with aphids.

At the end of the experiment Dr. Johnson extracted volatile chemicals from the bags and placed them in one of two chambers connected by a corridor. As expected, aphids placed in the chambers shunned the chamber with the chemicals produced by the infested plants. The aphids also shunned the chemicals produced by uninfested plants that had root and fungal contact with infested plants. Crucially, the aphids also shunned the chemicals produced by the uninfested plants that had a forty-micron mesh around them. This meant that the symbiotic fungi were not just supplying minerals, they were also relaying an SOS through the soil to other broad-bean plants. By comparison, the plants with the fungi-impeding half-micron mesh around them released chemical compounds that the aphids rather liked.

Enthralled by this, Dr. Johnson and his colleagues also ran tests on wasps. These are often found near bean plants infested by aphids because the wasps inject their eggs into aphids. These eggs rapidly mature, much like the beast in Ridley Scott's *Alien*, and then young wasps kill the aphids as they burst out of them. Astonishingly, the wasps were attracted to the very same chemicals that the aphids found repugnant.

The result seemed clear. Broad beans under attack were releasing chemicals that annoyed the herbivores eating them and summoned predators that would attack the herbivores on behalf of the plants. More important, they were using their underground alliance with the fungi to "tell" other broad beans in the area that danger was near so that they too would call out to predatory wasps for aid. As I read through all of this, I realized I had a surefire way to propose reporting on the paper to Geoff and get it printed for our 2 million subscribers: *Avatar*.

Using the blockbuster film as leverage, I argued that we should cover Dr. Johnson's research and point out that James Cameron's fictional moon of Pandora, with its network of plants that summoned the aid of the wild animals to fight back against the deforesting humans, was much more a matter of fact than fiction. As I expected, the article sailed through editorial and received a lot of positive feedback. It left me wondering, how did James Cameron know about this stuff? *Avatar* came out in 2009, and Dr. Johnson's work was first received by *Ecology Letters* only in November of 2012. Even the Chinese work that came before his was published only in 2010. Was Cameron moonlighting as a botanist? Or was he just lucky? As it turns out, he cheated.

If you search the *Avatar* credits, you will find the name Jodie Holt.[18] She's a botanist at UC Riverside and worked as a science adviser on *Avatar*. She is not a specialist on plant communication. Her area of expertise is how weeds develop resistance to herbicides, but she was definitely in a position to pass along information about upcoming research. Cameron wasn't lucky, he knew. He was well aware of where the field of botany was going and incorporated the ideas into his film. Yet that isn't the real genius of what he pulled off.

The aliens living on Pandora, known as the Na'vi, worship the plant network as a goddess. They don't actually understand the science behind how the network functions, but they value the results. They are in awe of their ability to tap into the network and download information from it in much the same way that I suspect the Vikings were in awe of the way they could be guided by ravens and wolves to food.

*Avatar* puts an anthropological spin on the subject. By exploring how scientists, greedy military types, and an alien species would interact with a complex plant network, Cameron challenges us to ask ourselves how we define something as magical. Is the network only magical to the Na'vi? Or do the human botanists, who learn how the network functions, come to view it in the same light? Does their knowing diminish the magic in some way?

## SPIDER SENSE . . . TINGLING

Oscar has an unusual job. Every day he surveys the patients in the rooms on the third floor of Steere House Nursing & Rehabilitation Center in Providence, Rhode Island. They are not in good shape. Most suffer from advanced dementia. Oscar's responsibility is to point out if anyone is facing imminent death so their loved ones can be brought in before the inevitable. It would be dismal work if Oscar were human. Fortunately, he is a gray, white, and fluffy cat, so it doesn't seem to be too much of a challenge.

Oscar simply wanders the halls, visits the patients one by one, sniffs the air in their room, and determines whether anyone is at death's door. If he decides that someone is, he curls up by the person's side and waits for the nursing staff to take notice.

Oscar came to my attention in 2007* when an article about his activities was published in the *New England Journal of Medicine*. That paper, written by David Dosa at Brown University, did something that no other research publication ever had. It brought me to tears. Here's an excerpt from Dr. Dosa's work to show why:

> *Oscar jumps onto her bed and again sniffs the air. He pauses to consider the situation, and then turns around twice before curling up beside Mrs. K. One hour passes. Oscar waits. A nurse walks into the room to check on her patient. She*

---

* He apparently came to everyone else's attention a few years later when his story was featured on an episode of *House*, which I've never watched but have been told is rather good.

*pauses to note Oscar's presence. Concerned, she hurriedly leaves the room and returns to her desk. She grabs Mrs. K.'s chart off the medical-records rack and begins to make phone calls. Within a half hour the family starts to arrive. Chairs are brought into the room where the relatives begin their vigil. The priest is called to deliver last rites. And still, Oscar has not budged, instead purring and gently nuzzling Mrs. K. A young grandson asks his mother, "What is the cat doing here?" The mother, fighting back tears, tells him, "He is here to help Grandma get to heaven." Thirty minutes later, Mrs. K. takes her last earthly breath.*[19]

To date, Oscar has presided over the deaths of more than thirty residents on the third floor. "There were a number of occasions where we had several failing patients at the same time and we didn't know who to go to first, but, invariably, Oscar knew. It was uncanny," explained Dr. Dosa. How Oscar does it is open to speculation. Dr. Dosa says, "It is possible that he is picking up on sweet-smelling organic compounds called ketones that get released by cells as they die."

Impressive as they are, it is important to take all of the findings surrounding Oscar with a pinch of salt. The cat has not, to date, been presented with any sort of controlled experimental conditions—for example, five patients are presented, one is truly dying, while the other four are unconscious but expected to remain alive for some time. Which does the cat select? I'm not suggesting that such a study would be feasible.* Casual observations of Oscar are probably the best we are ever going to get, but that isn't so in other areas of animal science.

Over the years, I have encountered dozens of studies suggesting that animals can help us better detect imminent danger from disease. In 2011, a team reported in the *European Respiratory Journal* that, while

---

* Or, for that matter, ethical. I can just imagine the letter of consent that you would have to write to a patient's family: "Hi, we noticed your grandmother is nearly dead. Would you mind if, on the day we think she is going to kick the bucket, we put her in an experiment to see if our cat can correctly predict that she is actually dying?"

lung cancer often lacks early warning signs, sniffer dogs can be taught to detect volatile organic compounds associated with the cancer.[20] The team ran a bunch of tests with both lung cancer patients and noncancer patients as controls to see if the sniffer dogs could reliably tease the two groups apart. To make things realistic, the team also included smokers who didn't have lung cancer and people with other respiratory problems, but not cancer, in their control group. The dogs correctly identified 71 of 100 cancer patients as having cancer, and 372 of the 400 control participants as not having cancer.*

A similar study was published in 2011 in *European Urology*, demonstrating that dogs could be trained to sniff urine samples and determine whether the urine came from a person who had prostate cancer.[21] Again the numbers were strong, with the dogs correctly identifying the presence of cancer in thirty of the thirty-three urine samples that had come from cancer patients. A 2012 study in the *British Medical Journal* presented analogous findings that were just as robust when dogs were trained to track down patients infected with the superbug *Clostridium difficile*.[22]

It isn't just death and disease that animals can detect long before we do. In the middle of the summer in 2003, a group of us were several days into a trek across one of the most forbidding regions of Romania, the Făgăraș Ridge. The days had been long, with distances covered averaging thirteen to fifteen miles. On the morning of the fourth day we broke camp and set out for a forest lodge on the other side of a few peaks. We wandered a pine-covered plateau for a few hours, and it was there that I first sensed something was wrong. Birds were flitting about everywhere, and squirrels were racing around, gathering food at a frenetic pace. The trees were buzzing with activity.

We reached a mountain hut, and with its offering of Romanian white chocolate cocoa and freshly baked pastries, many of us were keen

---

* I.e., there were only 28 false positives out of 400. Not too shabby, really.

on stopping for a bit. As we ate, the wildlife would not leave us alone. Robins and sparrows came swarming in to snatch every bite that they could, while the woodland rodents scurried at our feet to catch crumbs as they fell. Two hours later, after climbing up a peak on rusted cables while carrying fully loaded packs, we got hit by the storm from hell. Lightning lanced across the sky, thunder shook the stones beneath our feet and torrential rains came driving down. Within minutes we were soaked to the bone. Seven hours later, after battering our way through muddy thickets of monkey trees, scrambling down unstable scree slopes, fording a swollen river, and dodging motorcyclists on a narrow rural road, we made it to a lodge. During the train journey back to the airport in Hungary, I found myself staring at the notes in my journal that I had written just hours before the bad weather had come in: "Something is amiss. The squirrels are almost clawing at my feet and the birds are relentless. It is kind of like Snow White's menagerie, only all wrong."

I knew Snow White wasn't the only fictional fair maiden to have an army of forest friends keen to warn her when danger was near. Cinderella had her mice and doves, I vaguely remembered Sleeping Beauty having birds and rodents too, and I was sure there were other animated princesses that I wasn't thinking of. As I fell asleep on that train, I pondered whether there might be some kind of reality behind all of this. The idea just sat there in the back of my head, and then, nearly ten years later, a paper in *Experimental Biology* slid across my desk that made me realize there probably was.

The title was simple enough—"Do Birds Use Barometric Pressure to Predict Storms?"—but I was immediately wary. In my line of work, you all too often run into papers on appealing topics that have titles phrased as questions that end up yielding almost no interesting information whatsoever. Even so, my memory of the Romania fiasco drew me in and I read on.

The team, led by wildlife biologist Creagh Breuner at the University of Montana, started by making all of the logical arguments that I would expect in such a paper.[23] Mountain storms, particularly blizzards, pose a serious threat to small animals. Getting caught far from home could prove lethal to a bird if the wind started blowing fiercely and temperatures plunged. Being stranded in a nest with insufficient food reserves during a freezing multiday storm could be similarly harmful. Dr. Breuner's team also pointed out that numerous past studies suggested that birds had the ability to detect barometric pressure. Since barometric pressure typically plummets as a major storm moves toward a region, they speculated that birds might use their barometric sense to prepare for the worst.

To test this idea out, the researchers put a bunch of wild-caught white-crown sparrows in barometric chambers to simulate what an animal would experience as a major storm approached. To create a control group, some birds were monitored closely in chambers where no pressure change took place.

The researchers found that while the control group made an average of sixty pecks at their food dishes in four hours, the average for the birds exposed to the pressures mimicking an incoming storm was one hundred pecks. This suggested that, when sensing a storm, white-crown sparrows do increase their feeding.

A similar study published in *Current Biology* in December 2014 further supported these findings.[24] A team was testing whether tiny golden-winged warblers living in the Cumberland Mountains of Tennessee could carry geolocating technology on their backs. They could, but in the midst of the experiment, the birds fled their breeding site after having only recently arrived. Just a day later, a powerful super cell storm moved into the area and spawned eighty-four tornadoes that killed thirty-five people. Whether other small-bird species and rodents respond similarly is currently unknown, but I'd wager from my experience in Romania that they do.

Along the same lines, the Roman author Aelian, born in AD 175, wrote about animals fleeing before a terrible earthquake hit the ancient settlement of Helike:

> *For five days before Helike disappeared, all the mice and martens and snakes and centipedes and beetles and every other creature of that kind in the city left in a body by the road that leads to Keryneia. And the people of Helike seeing this happening were filled with amazement, but were unable to guess the reason. But after these creatures had departed, an earthquake occurred in the night; the city subsided; an immense wave flooded, and Helike disappeared.*[25]

This is one among many such stories that have been spun throughout history and are told to this day.

Remember the Asian tsunami that claimed over 220,000 lives on December 26, 2004? Many natives of Bang Koey, Thailand, suspected something terrible was coming on that morning when buffalo grazing near the beach suddenly stampeded through the village to a nearby hill.[26] Some took the stampede as a divine sign and followed the animals to higher ground. Others ignored it and were killed when the devastating wall of water arrived.[27] Was a deity temporarily using the beasts to lead faithful souls to safety? Were the buffalo sensing the shock waves from the far-off earthquake that caused the tsunami? Or were they sensing foreshocks? While I find the idea of a god leading his people to safety appealing, I'm inclined to believe the latter. Here's why.

## SOUNDS LIKE TROUBLE?

In 2000, psychologist Stanley Coren at the University of British Columbia ran a study to explore whether dogs suffered from winter blues, or seasonal affective disorder, as humans do.[28] He asked two hundred dog owners in Vancouver to report their animals' behaviors twice weekly throughout the year. He asked them to rate activity and anxiety levels on a 9-point scale. He found nothing of interest and, frustrated, put the study aside.

A year later, while glancing at the data once more, he spotted something. "While there were no trends that suggested the animals became depressed during the darker days of winter, there was this odd data point that I couldn't make sense of," Dr. Coren explained to me.

On February 27, 2001, 193 dogs, or 49 percent, showed high levels of anxiety.[29] Almost all the anxious dogs were more active that day as well. Dr. Coren assumed the spike in the data was the result of a passing thunderstorm or some other form of extreme weather, but as he scanned through Vancouver weather reports for that day, he found nothing of the sort. Then, as he went through newspaper archives, he noticed something intriguing. A 6.8 magnitude earthquake hit an area 240 kilometers (149 miles) south of Vancouver on February 28.[30]

For those who did not grow up in earthquake territory, a 6.8 quake is a major event that causes violent shaking in the area near the epicenter, has the capacity to bring down buildings, and can easily be felt hundreds of miles away. Were the dogs detecting some danger signals before the quake hit?

Dr. Coren wasn't sure, but he had read stories of avalanche rescue dogs hearing the cries of people buried beneath the snow that their human masters could not and wondered if any of the dogs in his study that were listed as deaf or hard of hearing had acted strangely during the day before the earthquake. Only one of the fourteen dogs with hearing troubles had acted strangely, and that dog was living in the same house as a dog that *did* have hearing and also acted with anxiety.

Geologists argue that Dr. Coren's results are just a coincidence because dogs could not possibly hear foreshocks from an earthquake nearly 150 miles away.[31] It is a fair point, but the statistical chances of Dr. Coren's findings being a fluke are less than one in a thousand. I'm inclined to believe there is something more to it given that science continues to reveal just how little we notice in the world around us that other species detect instantly.

Recently, I covered a bit of research that explored a rural area of New York. Known by Muggles as Jersey Hill and by homing-pigeon enthusiasts as the Birdmuda Triangle, this place has a long history of making even the most reliable pigeons vanish.

Why it caused the birds to disappear was a mystery for decades. Homing pigeons are well-known to make use of the earth's magnetic field as they fly, and the field was undisturbed in the area. Then studies revealed that the birds have the ability to detect something else that we humans are incapable of sensing: infrasound.

Created most commonly by ocean waves as they crash on the shore, infrasound travels for thousands of miles inland and plays a crucial role in helping pigeons to navigate their way home. Geologist Jon Hagstrum studied the Birdmuda Triangle intensively for infrasound and reported in *Experimental Biology* in 2013 that local atmospheric conditions associated with regional temperatures and wind directions were blocking infrasound in the area.[32] This blockage prevents homing pigeons from detecting a vital piece of navigational information and throws them off course.

So where does this leave Dr. Coren and his dogs? Well, pigeons form a far richer picture of the world than we do. They can navigate by the sun, detect magnetic fields, and hear infrasound. Given all of this, it seems both ridiculous and arrogant to argue that Dr. Coren's findings are mere happenstance until we have exhaustively determined what dogs do and do not sense in the world around them.

All of these findings made me wonder whether the people who came up with the "Snow White," "Cinderella," and "Sleeping Beauty" fairy tales had noticed small forest animals acting desperately before natural disasters and wove such observations into the myths as the "animals warning of danger" motif. Curious, I pored over early copies of these tales in the rare-books room at the British Library and discovered a lovely and gory side of "Cinderella" that I never knew existed.

In the 1812 version of the fairy tale that was collected by Jacob and Wilhelm Grimm, Cinderella is known as Aschenputtel.[33] She doesn't have a fairy godmother, but she does have a few doves who live above her mother's grave in a hazel tree. These birds speak with her and give her all of the nice gowns and slippers that she needs to attend the prince's ball. After she loses her golden slipper (it is golden, not glass, in the 1812 version), the prince visits her home with the slipper in hand to see if there are any maidens whose feet fit it. Aschenputtel's evil stepsisters' feet are too large for the slipper, so they both use knives to chop off various bits of their heels and toes to wedge them in.* The prince, being none too bright apparently, doesn't notice when they do this and takes one and then the other back to his palace for marriage. But as he passes the graveyard, the birds in the hazel tree call out:

> Backwards peep, backwards peep,
> There's blood upon the shoe;
> The shoe's too small, and she behind
> Is not the bride for you.[34]

Only after they call to him does he notice blood flowing out of the shoe and return to discover Aschenputtel. So that's kind of fun, but it isn't exactly a warning of a storm. Moreover, "Aschenputtel" is not the first incarnation of the Cinderella tale, just a well-known version written by the Brothers Grimm. The French author Charles Perrault created an earlier version called "Donkeyskin" in 1694 that lacks the doves.[35] A Chinese version of the tale, "Yeh-hsien," dates back to AD 850.[36] Again, it lacks the doves.** Furthermore, none of the early versions of "Snow White" or "Sleeping Beauty" have examples of animals pro-

---

* I did say *gory*.

** Although it does have a fish with red fins, golden eyes, and an amiable personality who befriends the Cinderella character and ultimately gets eaten by the wicked stepmother.

viding warnings of any sort.[37] One early version of "Little Red Riding Hood" has a cat that warns the little girl that her grandmother has been killed and that some wine she is about to drink is actually her grandmother's blood.[38] Nevertheless, that is just one version of many, and having a cat make such a warning is a far cry from a sparrow advising of a coming storm. So did the Disney people just make up this idea of animals warning their princess friends of danger? I am not sure.

While the classic fairy tales are thin on this kind of stuff, history is rich with stories of animals warning people of danger. The Roman historian Titus Livius reported that sacred geese living on the Capitolium in Rome warned soldiers of a Gallic attempt to take the citadel by surprise and stopped this catastrophe from occurring.[39] Lucius Cassius Dio, a Roman consul and historian, wrote that Boudicca, warrior queen of the Celts, traveled with a hare that helped her make military decisions.[40] Homer's *Iliad* describes Aphrodite as allegedly having sent doves to protect her son, the Trojan hero Aeneas, as he journeyed through the dark and dangerous forests at Avernus.*[41] The list goes on and on.

Did the Roman geese, Aphrodite's doves, and Boudicca's hare come from a realization that animals could predict impending death like Oscar, detect incoming storms like the birds in the *Experimental Biology* and *Current Biology* papers, or forecast earthquakes like Dr. Coren's dogs? We can't know that for certain, since we have no living Romans or Celts to interview. The argument that they may have just been making this stuff up always has to be given some consideration, but given how closely our ancestors looked for divine signals in animal behavior, I suspect that some of these beliefs were inspired by real-world observations.

---

* There is a good reason why birds would be excellent guides at Avernus, in southern Italy. More on that later when we discuss the dark oracle who took up shop at that site.

**FLIGHTS OF FANCY**

Aside from warning us of danger, animals, particularly birds, were used by our ancestors to predict future events. People would literally look into the sky and try to determine what would happen in the weeks and months ahead based upon what they saw. The questions that were being asked were basic. So too was the information being gleaned. It was almost always a binary question, with the messages sent by the gods through the birds being either positive or negative. Here is one example salvaged from an inscription found in the ancient city of Ephesus:

> *If a bird flying from right to left*
> *disappears, it is favorable;*
> *If a bird flying from left*
> *to right disappears on a straight course,*
> *it is unfavorable.*[42]

First, let's tackle the "left" and "right" descriptors in the above passage. The Homeric epics state that any person reading the birds must be facing north.[43] Thus, any bird flying to the right was flying eastward, and any bird flying to the left was traveling westward.[44] Birds appearing from the east, the direction of the rising sun, were considered a positive sign, while birds appearing from the west, the direction of the setting sun, were considered a negative one. Second, the practice of reading the birds was called οιωνός, which in ancient Greek meant both "omen" and "bird of prey."[45] Moreover, the most significant bird readings recorded in the ancient literature are associated with raptors such as eagles and falcons.[46]

So, from the look of things, our ancestors were monitoring the directional travel of birds of prey and using this information to make predictions. I found this interesting because, stored in the back of my head, I remembered a lecture during my first year in university on raptor biology.* Our professor found that during years when an El Niño or a La Niña system was present in the Pacific Ocean, the composition of the raptors seen from the local observatory changed.

What is El Niño? Well, it is a huge band of unusually warm ocean water that forms off the coast of South America every once in a while.[47] This warm-water band is almost always linked to high barometric pressure in the western Pacific Ocean and has a dramatic effect on weather patterns in South America and Asia. For example, during El Niño years South America tends to get hit with wet weather and floods, while southeastern Asia tends to suffer unusually hot and dry conditions. La Niña is the reverse. Linked to unusually cold water off the coast of South America, it brings low barometric pressure in the western Pacific Ocean and leads to drought in South America. It simultaneously plagues places such as Malaysia and Indonesia with floods.

I'd found it fascinating to learn that El Niño and La Niña altered the migratory paths of raptors. It also made perfect sense. As rainfall moved around, so did vegetation and prey. The raptors were just going where the food was. No big surprise. Similarly, raptors are lazy migrants. If they can avoid flapping their wings, they will. They do this by seeking out areas where the weather is conducive to their soaring along on columns of warm air rising up from the ground, called thermals. As regions with thermals shifted about, so too did the raptors' migrations.

If you were living in California and had no access to Internet, television, or radio, could you work out whether the world was in an El Niño or La Niña year by keeping an eye on which birds were migrating overhead? Yes. However, as I waded deeper into the literature on El

---

* Avian Biology 115. Quite a cool class. I got to handle eagles and great horned owls for course credit. I don't remember too much from the course, since I haven't handled a raptor in over twenty years, but I do recall two critical things. First, great horned owls have such strong talons that they crush your hand even when you're wearing a raptorproof glove. Second, you have to wear a fencing mask when handling eagles.

Niño and bird migrations, I discovered something much more intriguing. Monitoring bird migrations in relation to El Niño/La Niña might also be able to help us predict disease outbreaks.

In a 2013 paper published in *Proceedings of the National Academy of Sciences*, epidemiologists Jeffrey Shaman at Columbia University and Marc Lipsitch at Harvard University pointed out that the four most recent human influenza pandemics—i.e., those of 1918, 1957, 1968, and 2009—were all preceded by La Niña conditions in the Pacific Ocean.[48] They also noted the rich literature indicating that bird migrations were dramatically altered by El Niño and La Niña and suggested that these shifted migrations were fueling the formation of new influenza strains.

Dr. Shaman and Dr. Lipsitch argued that, because wild birds are considered the primary reservoir of influenza viruses that infect humanity, any major changes to their migrations could alter which birds and animals these viruses interact with. Weather patterns are changed all around the world by El Niño and La Niña, and these changes shift bird migrations everywhere. For example, during a La Niña year many bird species living in Asia interact with species, both birds and domestic animals, such as pigs, that they do not usually encounter. This increases viral mixing and paves the way for new nasty viruses to evolve. La Niña often increases the amount of floodwaters in Asia, potentially increasing the amount of viral transmission as bird and pig wastes get mixed together. Where the first cases of new influenza strains will appear in people after all of this happens is not easy to predict. A novel viral strain could jump from pigs to people locally in southeastern Asia; it could travel with the birds back to their summer ranges in Russia; or it could travel across several species and ultimately end up in African, Middle Eastern, or European towns. These ideas are relevant not only to influenza. Extensive literature documents the effects of El Niño and La Niña on outbreaks of other infectious diseases such as dengue fever, cholera, and malaria.[49]

Did our ancestors figure out that certain migrations indicated specific weather patterns were coming? Could they predict major disease outbreaks based upon whether specific raptor species were migrating east or west at certain times of the year? Armed with all of the migration information collected in North America and South America during past El Niño and La Niña years, I could certainly keep an eye on the skies on those continents, make my own predictions about pandemics, and stand a good chance of being accurate. What of the Greeks?

Researchers are only just starting to work out how El Niño and La Niña shape conditions in Europe. The continent has its own major climate system. Known as the North Atlantic Oscillation, it either sends lots of mild and wet weather slamming into Europe or shuts this weather off, leaving Europe with scorching summers and bitter winters. Even so, it looks as if El Niño makes northern European winters drier and colder than usual, while making winters in southern Europe milder and wetter.[50] Conversely, La Niña looks to cool the south and warm the north, but more work is needed for us to understand how these effects mix with the North Atlantic Oscillation system.

As for the journeys of raptors, three key raptor migratory routes head southward off the continent.[51] Known as the Palaearctic western, central, and eastern routes, most of these migration routes fork in Albania, Montenegro, and Macedonia. When they reach these points, birds have a choice. They can follow the western route out of Albania and fly over the boot of Italy through Sicily and into Tunisia.* Birds can take the central route south through Greece, over the Greek archipelago, and into Libya. Or they can take the eastern route out of Greece, through Turkey, Syria, and Israel and into Africa via Egypt.

---

* For all of you ornithology pedants out there, yes, I am aware of the more western branch of the Palaearctic route for raptors coming from areas such as France, Norway, and the Netherlands that goes through Spain and leaps over Gibraltar into Morocco. It just isn't relevant to making bird prophecies in ancient Greece.

The burning question is, do El Niño and La Niña determine which way a species goes in a given year? I have no idea and neither does anybody else. At least, not yet. However, based upon the literature that does exist, it seems a reasonable bet that they do.

Migrations are hard on birds and have a powerful evolutionary effect on populations through natural selection. Think about it this way: if you were a bird and you took "the scenic route" in a given year while all of your friends took a shortcut, you'd arrive at your southern location late, get last pick of wintering sites, get a poorer food selection, and be more exhausted from journeying farther.* All of this would leave you in poorer shape to breed and, in time, drive any genes that led you to take "the scenic route" out of the population.

Birds are finely tuned to know how to make their migrations as painless as possible, and I'm sure that if El Niño or La Niña makes one of the three Palaearctic routes more challenging, birds know about it before they depart and choose an easier option. As to which routes they might choose, my guess is that El Niño changes migrations little.

The evidence that we have is that El Niño warms the European south and increases the presence of thermals over Spain, Italy, and Sicily. If anything, El Niño makes the Palaearctic western route look particularly enticing and may draw birds from more central regions toward it. Indeed, in one 2011 paper on short-toed eagle migrations during 2010, which was an El Niño year, a team led by Ugo Mellone at the University of Alicante in Spain revealed in the *Journal of Avian Biology* that eagles tagged in Italy took a major detour, traveling hundreds of miles farther than necessary to fly over Spain on their way to Africa rather than just going over Sicily.[52] Was this because El Niño warmed Spain significantly and increased thermals such that it looked appealing

---

* Kind of like arriving at a party late and noticing that all the nibbles and champagne are gone, but ten times worse.

to migrants? We don't have enough evidence to support that hypothesis yet, but it is an interesting possibility. Yet it is with La Niña where things get fascinating.

If La Niña does cool southern Europe as climatologists suspect, this would reduce the presence of thermals in both Spain and Italy and drive migrations eastward.[53] Thus, if a bird species that normally headed west when it reached Albania was spotted taking an eastward migration route toward Turkey during a given year, this could be a sign that the world was experiencing a La Niña year.

Albania, Macedonia, and Montenegro aren't Greece. Most of Greece is too far south for anyone to see birds changing the Palaearctic route that they usually choose, but that is the Greece of today. The Greece of the past was considerably larger and included all of what is now Albania. Lots of Greeks in those areas were looking to the skies and trying to determine the will of Zeus by watching birds, and I'll bet that during years when La Niña was spawning pandemics, specific birds changed course and flew eastward, prophesying doom for all who saw them.

One of the most sacred sites in ancient Greece, suspected by some to have been founded as early as 2000 BC, is in a pretty good spot to watch which way birds are migrating. Known as Dodona, the site was sacred to Zeus and is close to the modern border with Albania.

Found near Epirus in Greece today, Dodona sits upon a rocky and arid hilltop and was home to an oracle who studied birds to make her prophecies.

# PROPHECY

## Liver Readers, "High" Priestesses, and Mind-Reading Devices

*Through the force, things you will see.*
*Other places. The future, the past, old friends long gone.*

—YODA, *THE EMPIRE STRIKES BACK*

Their voices filled the air with words of wisdom, guiding generals into battle and advising the droves on how to better meet their destinies. They were the oracles of ages past.

The writings of the poet Hesiod, who was born around 750 BC, and those of the historian Strabo, who was born in 64 BC, suggest that the prophecies made at the oracle of Dodona, a sacred site in Greece, were guided by the movements of doves.

Hesiod wrote in his *Catalogue of Women*:

*There is a land Hellopia with much glebe and rich meadows, and rich in flocks and shambling kine. There dwell men who have many sheep and many oxen, and they are in number past telling, tribes of mortal men. And there upon its border is built a city, Dodona; and Zeus loved it and appointed it to be his oracle, reverenced by men. . . . And they [the doves] lived in the hollow of an oak. From them men of earth carry away all kinds of prophecy—whosoever fares to that spot and questions the deathless god, and comes bringing gifts with good omens.*[1]

In his *Geography* text Strabo adds:

*According to the Geographer, a sacred oak tree is revered in Dodona, because it was thought to be the earliest plant created and the first to supply men with food. And the same writer also says in reference to the oracular doves there, as they are called, that the doves are observed for the purposes of augury, just as there were some seers who divined from ravens.*[2]

The trouble with Hesiod's and Strabo's reports is that they were both written more than a thousand years after Dodona started functioning as an oracle, and they do not give us a clear image of what led people there in the first place. Was it really doves? Or were people originally looking at raptors and then later looking at doves? Initially, I assumed this had to be the case, because doves don't migrate, right?* Well, I was wrong.

Some doves do migrate, such as the European turtledove. It spends its summers in Europe and its winters in Africa. How does it make its way south? The species takes all three of the Palaearctic routes that I mentioned in the last chapter and funnels in greater numbers down the various routes from year to year depending upon conditions.[3] Was this what the oracle at Dodona was initially looking at? Maybe, but the oracle in later years made divinations by studying the movements of leaves in the oak tree at the site and listening to the reverberations of bronze cauldrons that were placed in a circle around that oak.[4] Could these too have helped her to make predictions about the future? I doubt it. My suspicion is that as Dodona grew in fame, it became more important for the priestesses of Zeus to put on a good show so they could better collect donations from the pilgrims who visited. Whether they were still looking at the migrations of birds is a detail that has been lost to the ravages of time.

---

* In Avian Biology 115, aka Raptor Biology, we readily dismissed all doves seen in the field as "falcon fodder," which should give you an appreciation for just how much attention we were giving them.

As I turned my attention to an area about five hundred kilometers west of Cairo, I found descriptions of the lush oasis of Amun, the king of the ancient Egyptian gods. There, priests would look upon the god's image hung in the shrine and watch its movements to determine the deity's responses to questions.[5]

Admittedly, it does sound a little crazy. Making major decisions based upon how the image of a god reacted to invisible currents of air in a shrine? Our ancestors couldn't have been that gullible, could they?

The oasis at Siwa, where the oracle of Amun was based, is nothing short of a miracle on earth. Past the lush palms and the emerald pools of the oasis sits Egypt's scorching and desolate desert. Just as the Mormon pioneers of the American West concluded that the stunning cliffs of Zion, Bryce, and the Grand Canyon were sacred places, it is easy to look at the verdant majesty of the oasis and conclude that Amun had touched the place. Pour in a bit of creative thinking, and the idea of it pulsing with mystical power gets less crazy by the minute.

While not much of the shrine remains, and we have little physical description of it by ancient authors, this was a majestic site dedicated to the king of gods. Even Alexander the Great visited the oasis, in 331 BC. He was not the sort to visit just any old oracle.[6]

Alexander was granted a private audience, something that was almost never done. As a result, his questions, and the god's answers, were never recorded. Even so, many speculate that he was asking whether he had Amun's blessing to invade the rest of Egypt and got a positive response that drove him to continue his historic conquest.[7] He later asked to be buried at the oasis, driving historians to suspect that he quite liked what the oracle at Amun had said to him, but nobody knows.

However, oracles were not limited to Egypt and Greece. Some made their way as far west as Britain, with a few individuals taking up shop at the sacred spring in Bath. Most people wandering the Roman ruins there today come away well aware that this was a place of relaxation. A handful who take the time to more closely study the site realize that it was also a healing sanctuary. Few realize that it was also a place of prophecy.

Given the impressive nature of the green bubbling springs, I expected to find that thermal waters played a major role in the prophesying that took place at Bath. Not so. The oracles there were entirely focused upon guts.

The priests would slice open animal sacrifices, pour their insides across a table, and then look at how the juices and organ bits had spilled out to make their predictions. Yeah, I know, it sounds like rather gross stuff. However, the god at Bath was not a grim deity of death but a rather odd mix of Athena, the goddess of wisdom, and the mysterious Celtic nurturing mother goddess, Sulis. This was a place of light, knowledge, and health. How animal-gut analysis got into the picture is not known for certain, but we can speculate.

Studying animal guts for signs of what might happen in the future was widespread in the Roman world. If you were reasonably well educated, good at making guesses, and comfortable with gore, setting up shop at a mystical site was a sensible business move. The impressive backdrop of Bath would only enhance the perceived power of prophecies. It would not make an aspiring prophet any more "right," but it would increase patron belief and draw in more donations, which was what mattered. Even so, as I read further about the practice, I started to wonder if there was actually something to it.

## A LOVE FOR LIVER

While predicting the future with guts was common in the Roman Empire, the Romans didn't come up with the idea. The Greeks read entrails before the Romans, and even earlier were the Hittites, the Babylonians, and the Assyrians.[8] Using entrails to predict the future goes back nearly three millennia in Mesopotamia.[9]

In its earliest form, entrails reading was much more focused on a single organ, the liver. Readings were fundamentally a binary system, a simple yes or no. How these answers were obtained is not entirely clear, but Derek Collins, a classicist at the University of Michigan, argued in a paper in the *American Journal of Philology* in 2008 that a mix of features of the liver was studied.[10]

After combing through the literature, Dr. Collins worked out that a dense liver was a good sign. A bright color and smooth texture were also positive. In contrast, a blocked portal vein or a bent caudate lobe was a bad sign. It was even worse if the caudate lobe was entirely absent, which indicated that something terrible was going to strike a region or country as opposed to just a single person or family. However, the most intriguing thing that I found in Dr. Collins's work was a bit of text that he translated from Cicero's 44 BC philosophical treatise, *De divinatione*. Democritus, the ancient Greek philosopher widely considered to be one of the earliest scientists, explains:

*The ancients widely established that the entrails of sacrificed animals were inspected because the general condition and the color are sometimes prophetic of health, sometimes of sickness, and sometimes also of whether there will be sterility of the fields.*[11]

Dr. Collins explained that when ancient towns were about to be founded, liver readers, or hepatomancers as they were known, would be called on to kill an animal grazing in the area and predict whether it was a good place to build homes and establish farms.[12]

Were our ancestors looking at livers to determine whether an area that they were thinking about developing was subject to episodes of drought or disease? Could livers store this information? Curious, I contacted Dr. Collins and asked if he had ever considered the possibility that something real might be behind liver prophecy.

"I've long thought that the liver could record the environment in some way, but being a classicist and not a veterinarian, I really have no idea," explained Dr. Collins. As a paleontologist by training, I only knew that bones were capable of recording some diseases.* I wasn't sure whether any sort of medical history could be found in the liver. It was time to have a word with a doctor of veterinary medicine.

Bruce Rideout is no ordinary veterinarian. He's the director of the Wildlife Disease Laboratories at the San Diego Zoo and a researcher whom I have been in contact with since my earliest days as a science journalist back in 2000. If anyone on the planet could advise on what, if anything, a liver could say about an animal's environment, it was going to be him. So I gave Dr. Rideout a call and explained my rather unusual question.

"You'd definitely see evidence of an animal's current nutritional state if you examined its liver. You might even see some evidence of past diseases," he explained.

Aware of the historical literature, I pressed for specifics.

"Well, a good and healthy animal would have a firm, medium-brown or mahogany-brown liver with uniform consistency, sharp edges, and no lesions. An animal that had been healthy but had suddenly been forced to metabolize fat [because of a famine or some other disaster] would have a soft, yellow-tan liver with rounded edges." Dr. Rideout knew his livers, and his descriptions hinted that color was a critical factor just as our ancestors had suggested, but I still wanted to know more. There was only one way forward: I was going to have to get my hands dirty.

"Yeah, hi, my name's Matt Kaplan. I'm a science correspondent at the *Economist* and an author working on a book about the science behind historical magic practices. I was wondering . . . do you have any spare livers that I might be able to study?"

---

* Paleopathology—the study of illness in fossils—is getting to be an unbelievably cool field, providing the recent discoveries that ancient rhinos suffered from chronic arthritis and that dinosaurs had all kinds of degenerative diseases.

As I called London butchers with this question and was met with a mix of dismay, disbelief, and, in one case, a lot of swearing, I started to wonder if I was out of my mind. Even so, I ultimately lucked out with Provenance Butcher in Notting Hill. The fellow on the phone said, "Hang on a moment, I've got just the man for you to talk to."

So I met Struan. Struan had been in advertising for years, got bored with it, took a crash course in butchery, and started his own store. "Oh, yeah, the life an animal leads definitely would show up on its liver. I'm pretty sure you'll be able to readily see it," he said.

This, of course, led to my next question: "Would you mind if I came over to examine a few livers sourced from different environments?"

The line went silent for a moment and my stomach went into a knot. "Sure, that sounds like fun!"

My editor, Geoff, who studied zoology at Oxford before turning to a life of science journalism, also thought the activity sounded like a laugh. Thus, when the day arrived for me to visit the butcher, he tagged along.

"One of these livers came from a pig that never saw the light of day and lived in horribly cramped conditions in Denmark. The other comes from a British pig that was able to roam around outdoors and live a pretty good life before it was slaughtered," explained Struan as he put the two pieces of meat on the work surface.

"Well, they certainly do look different," said Geoff as he reached out to handle one of the livers. "What is it we're looking for, mate?" he asked me.

"Color, density, orientation of the caudate lobe, smoothness of the portal vein," I read off from my notes. We worked at it together for ten minutes while studying liver diagrams that we'd brought with us. Then we made our predictions to Struan. "This one is much denser, more colorful, and smoother than that one, so it must be the liver of the farm-reared pig, and that pale and soft thing over there must be the liver of the pig that lived the miserable life," I said with confidence.

We'd gotten it absolutely wrong. Then, when it came to looking at chicken livers, we proved just as useless.* "In my experience, the darker the liver, the harder the organ was having to work and the lower the quality of the animal's life," explained Struan. When I mentioned Dr. Rideout's suggestion that an unhealthy liver would look yellow-tan and be quite soft, Struan smiled and said, "You gotta remember, the vets at the slaughterhouse don't let the meat from even slightly ill animals reach us."

It made perfect sense. Struan never saw livers with parasites in them, nor did he ever encounter the livers of animals suffering from serious diseases or that survived famines. Those animals and their tissues were destroyed long before they could ever make it to a butcher in London. Even so, the experience proved valuable.

Both the intensively raised and the farm-reared pigs were healthy animals that were suitable for human consumption. Yes, one would have led a life that we would call happy and the other unhappy, but both would have had plenty of food and spent their days free of most diseases. However, even with these similarities, their livers looked staggeringly different. They were different enough for me to now be able to tell the two apart without much trouble. Diseased livers and the livers of animals experiencing hard times, Struan advised, would look more different still. "I'm pretty sure we'd even be able to get a sense of general potassium and iron levels in the fields where animals were grazing based upon liver characteristics," he said.

As I sat down at a pub with Geoff for a meal afterward,** he said, "You know, it's pretty well documented that if you make a pig's life nice, it increases yields." He also explained that stress released lactic acid into the

---

* We would have looked at cow livers too, but those were out of my price range. Remember, people like cow liver, so it is expensive. In the absence of amateur diviners, chicken and pig livers are usually just thrown out.

** He ordered pork. I just couldn't bring myself to do so.

flesh and had the potential to ruin meat. It was all intriguing, if not a bit unnerving, as the prosciutto hit the table. This all connected to the Greek author Philostratus. He wrote that sacrificial animals should have no knowledge of their coming death, because if they did, their livers would get frightened and the prophetic messages held within would be damaged.[13]

In fact, pig liver, I discovered after my experience at Provenance Butcher, was generally not considered a good liver to look at if you had cattle or sheep to kill instead. Pig livers were bad for divination, the classical literature argues, because pigs have too much "spirit" in them.[14] Precisely what this meant, I wasn't sure, but when I mentioned it to Dr. Rideout, he laughed and said, "Well, male pigs have an incredibly strong odor in their skin that you cannot get off your hands for days, no matter how much you wash. Maybe that's why they thought they were no good to work with!" I laughed too, and for a moment we both lapsed into silence. "You know, there might be something else to it," Dr. Rideout added thoughtfully. "During hard times, pigs will eat just about anything. They are amazing survivors. If food is scarce and there's a carcass rotting on the ground, they'll eat it. Cattle and sheep won't do that. Perhaps that opportunism made their livers less useful when people wanted to glean information about famines and droughts."

As our conversation was coming to a close, a final question came to my mind that I figured only Dr. Rideout could answer: "I'm guessing this is probably going to be news to you, but making prophecies from livers was not unique to the Old World. The Inca studied llama livers, and the Aztecs gazed at pelican livers, because they viewed pelicans as the lords of the seabirds.[15] I'm just curious—might studying pelican livers have informed the Aztecs about what sort of seafood harvest they could expect in a given year?"

Dr. Rideout thought the matter over aloud for a moment. "Well, you've got El Niño and the Pacific Decadal Oscillation System . . .

Both would push fish populations around, and if fish populations moved away, pelicans would suffer. Studying pelican health may have been able to tell the Aztecs a fair bit about what the year ahead had in store for them."*

I am not going to argue for a moment that our ancestors were always wielding livers intelligently. Indeed, when Hannibal, the Carthaginian general famous for trampling Romans with elephants, was living in sanctuary at the court of King Prusias, he was advised by the monarch that war was not an option because the entrails were forbidding it. When met with such stupidity, Hannibal replied, "And do you believe a few pieces of calf meat more than a veteran commander?"[16] In spite of this, I think there is enough evidence for us to suspect that these practices were at least sometimes based upon a reasonable degree of rational analysis of the surrounding world. Of course, the power wielded by the liver prophets was nothing compared to the most powerful oracle of them all.

## THE PREDICTION MACHINE

Her prophecies were everything. They drove city-states to war, determined the fates of kings, and guided the development of the earliest democracies. People traveled hundreds of miles to hear her words of wisdom and make sacrifices to her god, Apollo. No oracle anywhere else in the world was viewed with the same level of respect or listened to as closely as the oracle of Delphi. She was, quite literally, a living wonder of the world.

Every eight years in Sparta senior officials would gaze up at the sky, and if they saw a shooting star moving in a specific direction, they deemed it a sign that the gods were angry at their two kings for behaving badly.[17] However, the Spartans respected the oracle of Delphi so completely that they granted her the power to intervene. Their laws

---

* Being me, I also asked if he could collect a mix of sickly and healthy pelican livers that I could study. He said he'd look into it, but I'm not holding my breath.

dictated that she, and she alone, had the ability to disregard the celestial omens and keep their monarchs in power.[18]

The people of Athens were similarly guided, such as when, in the sixth century BC, the legislator Solon came to Delphi to ask for legal advice. He was developing reforms for Athenian society, and the oracle is reported to have advised him, "Take thy seat amidships," as he wrote his laws.[19] These words are credited with inspiring Solon to construct civil codes that dealt equally with both the rich and the poor. Solon certainly seems to have had great respect for Delphi because he put the oracle in charge of selecting the men in Athens who were responsible for interpreting sacred laws.[20]

Delphi was always consulted during emergencies. When the Greeks were attacked by Persia, the oracle told the Spartans that, if they were to survive, they would need to lose either their town or a king. This proved true. In the epic battle at Thermopylae, the Spartan king Leonidas died in combat with all his men. This then paved the way for the Persians to march deeper into Greece, where they ultimately met their doom at the hands of the Athenians.[21] Of course, the Athenians had Delphic advice of their own.

In the days before the vast Persian army came crashing down on them, the oracle prophesied that a great wall of wood was to be their salvation. According to the historian Herodotus, the Athenians took this to mean that they should fight the Persians at sea on their wooden warships.[22] This move, legend says, saved them, as they used their boats to outflank the Persian ships and sink them in a narrow passage between nearby islands.

So respected was Delphi that in the middle of the sixth century BC, when the site caught fire and burned down, contributions came in from throughout the classical world to organize a rebuilding effort and construct a much larger temple complex.[23]

All of these stories raised questions in my mind. If the oracle at Delphi had experienced such fame for just a single generation, I could shrug off her prophecies as the mutterings of a lucky individual, a knowledgeable woman, or a little bit of both, yet the site of Delphi held tremendous power for almost a thousand years. How could that be possible?

A skeptic at heart, I found Delphi presented me with an intriguing problem. While the site itself remained the same throughout the ages, the oracle, a high priestess of Apollo called the Pythia, changed over time. Most history books report that as one Pythia died, a new one was selected from the nearby community of Delphi. Nevertheless, as I looked closer, it became obvious that this was not how things worked.

It turns out that the Pythia was not one but many. A secret sisterhood of sorts existed, and members were selected by a process shrouded in mystery.[24] Whether potential oracle candidates had their intelligence or political awareness tested in some way is unknown, but their careers did overlap. At the height of its fame, there were three Pythias working in unison, with two always on call for delivering prophecies and one standing by as sort of a pinch hitter if one of the others fell ill.[25] As much as we do not know about these women, one thing is certain: these priestesses were in a remarkably good position to become well educated.

People came from far and wide to present their questions to the oracle, and before they were allowed access, they were extensively interviewed by the priests and priestesses working at Delphi.[26] A perk of this job was to learn a great deal about matters of religion and state in the surrounding world. The clergy at the temple would have grown knowledgeable and been in an excellent position to increase the chances that their prophecies would come true. The evidence we have hints that the oracles at Delphi got things right more often than not because the temple was a well-oiled machine for intelligence gathering and wielded it to guide their predictions.

## THE HIGH PRIESTESS

However, even if I am right and this is how Delphi functioned, a lot of unanswered questions remain about the place. As solid as the prediction-making machine might have been, this system would have depended upon a constant flow of information from pilgrims who would have journeyed there only *after* the site had become popular. So how did Delphi garner its reputation to begin with?

Before giving out prophecies, the Pythia took ritual baths and drank from springs that were thought to be guarded by sacred spirits.[27] Once properly purified, she went to a place deep within the temple called the adyton. Inside this small, sunken room, she seated herself on a tripod and inhaled a gas rising from what was either a natural fissure in the ground or a spring.[28] Shortly thereafter, she entered a trance and answered questions that were presented to her by clergy and important guests sitting just beyond the sacred chamber.[29]

Visitors, ranging from commoners to scholars such as Strabo, Aristotle, and Herodotus, report a sweet, perfumelike scent in the adyton, which they believed to be the odor of Apollo's spirit as it passed through the Pythia's body.[30] But subsequent scholars suspected otherwise.

Plutarch, a high priest at Delphi during the end of the first century AD, noted that during most sessions the Pythia appeared to be mildly intoxicated while still being able to recognize visitors standing nearby, comprehend questions, and respond in poetic verse. Plutarch also wrote that on some occasions the Pythia foamed at the mouth, and when pressured by a wealthy patron to go into the adyton when she did not want to, she became violently delirious and died.[31] Plutarch speculated that these emissions were connected to geology.

By the time Plutarch was supervising religious activities at the site, the oracle had already experienced a steep decline in power. While she had once given clear answers, she was now providing prophecies that either had double meanings or were blatant riddles. This change led Plutarch to question if the flow of the sweet-smelling gas in the adyton might have been disrupted by movements in the earth over the years. More specifically, he wondered if an earthquake that had struck the area in 373 BC might have been responsible.[32] He died before he could ever explore this question, but he was probably right.

At the very end of the nineteenth century, a French team of archaeologists came to excavate Delphi. Well aware of Plutarch's writings, they expected to find an obvious fissure in the adyton with sweet-smelling gases blowing out of it. After years of hard work, they did not find the remains of the adyton and saw no evidence of a gas-producing fissure. In 1927, they were forced to report that Plutarch's speculations were erroneous.[33]

For decades Delphi was left alone, then along came a geologist from Wesleyan University in Connecticut. Jelle de Boer was advising the Greek government on where it would not be safe to build a proposed nuclear power plant, so he was looking all over the mainland for active fault lines. As he drove past Delphi, his geologically trained eyes immediately spotted an active fault running right underneath the oracle's temple.[34] Later, during a return visit, he spotted another.

Dr. de Boer had read Plutarch's writings and guessed that the priest had probably been correct when he speculated about emissions being related to geology. With faults crossing one another in an X formation beneath the temple, it seemed likely that some gases may once have been spewed out. When he communicated his thoughts to archaeologist John Hale, at the University of Louisville, he immediately met with resistance.[35]

Dr. Hale, well schooled on the French excavations at Delphi, knew that gas-producing fissures had never been found in the adyton.[36] However, as Dr. de Boer pressed the matter, Dr. Hale realized an important detail. He didn't know what a fault would look like.* Archaeologists were not taught much geology at university, and this led Dr. Hale to question if the French team had gotten things wrong.

---

* The full and engaging tale of Dr. de Boer and Dr. Hale's interactions are reported in William Broad's *The Oracle: Ancient Delphi and the Science Behind Its Lost Secrets*. Well worth a read.

Upon closer examination, Dr. Hale realized the French had assumed that the only geologic features capable of producing gas emissions were volcanic. When they found none, they threw their hands up in defeat* and walked away. But the pressure created by tectonic plates pushing up against each other, as they do at Delphi, can also generate enough energy to heat water in the ground.[37] This water rises through fissures and carries with it chemicals from the surrounding rocks. As Dr. de Boer ran a geologic analysis of the Delphi site, he found evidence of petrochemicals in the limestone rocks below the temple. He theorized that hot water had once been generated by tectonic activity and carried petroleum-based compounds such as methane, ethane, and ethylene up to the surface, where they were inhaled by the Pythia.[38]

Dr. de Boer and Dr. Hale got permission from the Greek government to search the adyton for evidence of mineral deposits. They guessed that if spring water had once been flowing out of limestone rocks in the area, then it would have left behind limestone crusts, similar to stalactites and stalagmites. Furthermore, they expected that if any gases had bubbled out in large quantities, then some of them might have become trapped in the crusts as they formed.

Sure enough, the scientists found and collected limestone crusts below the temple. They also collected water samples from nearby springs that likely once flowed beneath the oracle's site. All of the samples were sent to a lab for analysis, and the chemistry confirmed what they had been thinking: petrochemicals had been bubbling up.[39]

They reported in August 2001 in the journal *Geology* that the limestone crusts had captured the gases ethane and methane when they formed thousands of years ago, and the spring water was still releasing small amounts of ethylene.[40] None of this could confirm exactly which

---

* As a British citizen, it is my duty to point out that this is something that comes rather naturally to the French.

vapors the Pythias had been inhaling, but ethylene smells sweet like perfume and, when breathed in low concentrations, has some pretty profound effects on the body.

"It's an anesthetic," explained toxicologist Alan Woolf at Harvard Medical School when I asked him about it. "Unfortunately, it's also explosive. One spark in the operating room with it present, and *boom*. It hasn't been used in that way for a long time." Could the Pythia have been thinking rationally while breathing concentrated ethylene? Dr. Woolf swirled his coffee around in his cup for a moment while mulling over the matter. "I'll bet you can. You would still have your faculties, and you definitely wouldn't be comatose. You'd just be feeling really, really great," he explained. The relaxation granted by the gases might have allowed an expert Pythia to more effectively consider political matters without being distracted by unwanted surface thoughts. Was that the case? I like to think so, but I doubt we will ever really know.

To prophesy is to peer into the minds of gods, whether by looking at migrating birds, dissecting livers, or studying the words of intoxicated priestesses. Yet, a flip side to prophesy is just as intriguing and medically more important: peering into the minds of people.

## THE XAVIER DEVICE

As Charles Xavier in the *X-Men* films, James McAvoy just touches his temple with his fingers to suggest that he is reaching into someone else's mind. Patrick Stewart, his predecessor in the role, did even less. He merely squinted his eyes a bit as he used his power. Jedi mind reading in *Star Wars* is much the same. There are no special effects, just a pregnant pause as Darth Vader reaches out with his mind to learn that Luke Skywalker has a twin sister.

Being able to read minds would be awfully handy. In the world of law, peering into someone's thoughts and determining whether they are telling the truth would reduce the length of a great many court battles, saving everyone* a whole lot of money. You could argue that we already have this ability with lie detection tests, but most courtrooms won't even admit their results as evidence. This makes sense, because the polygraph, as the lie detector is called, only measures thoughts indirectly by monitoring things such as breathing and blood pressure.[41] Sure, when someone's respiration rate and blood pressure increase, this hints that they are more anxious about what they are saying and suggests they could be lying, but they could also just be nervous about telling the truth. For this reason, it would be better if brain activity could be monitored directly.

Using functional magnetic resonance imaging (fMRI), research groups around the world have been exploring how much oxygen different bits of the brain consume when presented with different tasks.[42] This method allows brain activity to be monitored closely and, for those interested in detecting liars, has yielded some interesting results. Certain areas of the brain—the left prefrontal cortex and the anterior cingulate cortex—activate when someone is lying but not when the person is telling the truth.[43] A lot of work remains before fMRI readings can be used in the courts, but I suspect they will be one day. Even so, this work pales in comparison to what other teams are accomplishing with fMRI.

If losing one's mental faculties is a tragedy, then losing control of one's body while still having a fully functional mind is a horror. It is, blessedly, not all that common, but some diseases leave patients completely paralyzed. They cannot move their arms, hands, eyes, or mouths, yet their brains still appear to be behaving normally. Often called "prisoners in their own bodies," these patients are in a terrible position. They can think but cannot communicate their feelings, needs, or wants. Finding a way to figure out what is on the minds of these patients would be valuable.

In a study led by Marcel Just at Carnegie Mellon University, par-

---

* Minus the lawyers.

ticipants were given specific places or things to think about. Each of these focal points fell into one of twelve categories. Using fMRI devices, researchers were able to correctly guess which category was on a participant's mind more than 80 percent of the time.[44]

One research study, led by Brian Pasley at the University of California, Berkeley, tried to work out what word a participant was told by monitoring electrical activity in the auditory cortex of the brain. Some words were real, such as *jazz*, *deep*, and *cause*, and some were pseudowords, such as *fook* and *nim*.[45]

The team designed a computational model that produced sounds based upon the areas of the cortex that each word activated. The researchers would listen to the sounds created by their computers and guess what word the sounds represented. They reported in the journal *PloS Biology* in 2012 that they were almost always able to correctly guess the words.[46]

All of this research has a long way to go, but I still find myself marveling at how quickly fiction is becoming reality. While we may not possess the mind-reading powers of Charles Xavier, we may one day see devices that allow us to read the thoughts of those around us. Whether such inventions will be viewed as magic made real or the stuff of the darkest science fiction will, I suspect, depend not so much upon the state of the technology as upon the state of human ethics when the technology becomes readily available.

On that cheerful note, let us turn to the darkest supernatural forces of all, those associated with the underworld.

# BEYOND THE GRAVE

## Underworld Gateways, Guardians
## of the Dead, and Near Death

*DON'T THINK OF IT AS DYING. JUST THINK*

*OF IT AS LEAVING EARLY TO AVOID THE RUSH.*

—DEATH, *GOOD OMENS*

In Italy, only a short distance north and west of modern-day Naples, lies the community of Cumae. While quiet now, it was bustling during the days of the Romans. Littered with temples and catacombs, Cumae held considerable religious significance. It is easy to see why.

On the side of a lush hill, at a shrine dedicated to Artemis, the goddess of the moon, ancient followers would have seen the full moon rise perfectly between the columns of the shrine's portal on specific holy nights. The shrine of Apollo was similarly built to allow first glimpses of the rising sun over nearby lakes. The temple of Zeus sat above all the rest with commanding views of the landscape as far as the eye could see.

The shrines and temples are impressive, but as I walked the grounds, I came upon something beneath them all that seemed very odd: an oracle dedicated to Apollo. The oracle's shrine is about as dark as any could be. The Italians have installed electric lamps, and with a bit of imagination, you get a sense of what it would have once looked like.

Carved into the side of the long cave are arches that face westward.

They allow just a bit of afternoon sun to trickle in. My footfalls echoed ominously, bouncing off the alcoves like an acoustic form of vertigo. During the day the oracle's cave was a bit spooky; at night it would be downright terrifying.

None of this made any sense. Why would an oracle working for the god of the sun, light, youth, and prophecy work in such a dark and terrible cave? The arches did not even face east to catch the morning light. At best, they would have caught the rust-red setting sun and made the site even more frightening. I knew pieces of the puzzle had to be missing, and as I explored the ruins, I found my answer in the form of a 3.2-inch (8.2-centimeter) disk excavated decades ago.[1]

In a script that was most common during the seventh century BC, the disk read, "Hera does not permit you to return and consult the oracle again." The content was almost comical. Nearly three thousand years ago the oracle working at Cumae had put up a sign that was effectively saying, "If your prophecy doesn't come true, you are not welcome back to give me trouble!" It made good business sense. The oracle would not want people to return and scare off other potential patrons. It would generate bad publicity and hurt donations. Amusing as this was, a far more important piece of information was on the disk. The god associated with the oracle in the seventh century BC was Hera and not Apollo.

Hera, the jealous wife of Zeus, was the goddess of fertility. But Hera's divine portfolio was not always just associated with bringing life into our world. She also once held sway over the forces that shepherded life out of it. She was originally the goddess of shades and souls lingering just beyond the fabric of the mortal realm.

Things were starting to make sense. The grim site of the oracle had not initially been connected to Apollo at all; that came a lot later.[2] The people who founded this oracle site were Hera worshippers, individuals keen on speaking with the spirits of recently departed loved ones. That was why the cave was so filled with dread and why none of the dawn's light would ever filter in. Apollo was brought in once Hera's following had shrunk.[3] The decision to toss out Hera and install Apollo was an economic one.

Electric with enthusiasm upon realizing the history behind the oracle's site, I searched for more information and learned that the ruins on the hill were not where religious activity at Cumae started. The true origins of the site were about a half mile away on the edge of a lake called Avernus on my map. That name seemed familiar when I first read it, but I couldn't work out why.

My time at Cumae was limited, so I told my traveling companion, Charlie, who was just tagging along to see what one of my "business" trips was like, that we needed to leave the ruins immediately. We raced down to the lake and were immediately disappointed. Not a single ruin was in sight.

While driving around the lake, we found a large cave nearby. Annoyingly, it was blocked off by a thick metal gate. The cave was dark and grim just like any other, with a cold wind blowing out. That meant air was entering an opening to the cave somewhere else. Could a pocket somewhere deep inside have been the initial site of Hera's oracle? I wanted to get inside, but found nowhere to squeeze past the gate.

Frustrated, we scouted more locations around the lake and soon came upon another clue in the form of a small wooden sign that read IL BANE DI SYBIL DI CUMA, which meant "the bath of the oracle of Cumae." Intrigued, we followed the path adjacent to it.

While we were a good fifteen feet above the water level of the lake, the trail was saturated and surrounded by dense vegetation. Groundwater was traveling through the rocks and making its way to the lake through the soil beneath our feet. Moreover, the sign spoke of a bath of the oracle. This sent my mind racing. Could a natural spring be hidden in a grove? I certainly hoped so. Then I saw it, another friggin' gate bolted to solid stone in front of a cave. After swearing several times at the Italian government for taking all the fun out of my day, I looked more closely. This gate was not just a solid mesh of metal, it was a door. A door with a lock. I looked over at my traveling companion.

Charlie was truly just along for the ride, but he was no ordinary tourist. I tend to befriend eccentrics, and Charlie was no exception. Yes, by day he was a computer programmer, but by night he was a scholar of ancient history, a black belt in aikido, a masterful juggler, and, most important, a talented lockpick. He knew exactly what I was thinking and said what I didn't want to hear: "Sorry, Matt, I left my lockpicks in San Francisco."

I was disappointed, but I couldn't fault him. Authorities in most parts of the world do not look kindly upon lockpicks found in baggage. Chalking the matter up to experience, I pressed my face against the iron bars and shined my light inside.

The air was thick with moisture. Moss grew heavily along the nearby walls, and cave popcorn speckled the walls. Like stalactites or stalagmites, cave popcorn forms as water rich in calcium carbonate splashes against stone. The formations were unmistakable. This place had seen a lot of water once upon a time.

As we were leaving, I figured I should, at the very least, take a water sample from the lake for analysis. Perhaps the lake's water had some unusual chemical properties that ancient locals might have attributed to magic. Spotting a small pier just a few hundred yards away, we drove over. I told Charlie to wait in the car, as I would be just a minute, but he got out anyway to look around. It was a good thing he did.

As I knelt down to collect the water, he interrupted me and pointed to a sign just above my head: IL LAGO OCCUPA UN CRATERE VULCANICO. The discovery was as awesome as it was logical. The area around Naples is geologically active. The Roman cities of Pompeii and Herculaneum were destroyed by the eruption of Mount Vesuvius, but Vesuvius is not the only volcano in the area, just the biggest. In moments I was on the phone with a geology friend back at Imperial College in London.

Like so many of my conversations, this one started with the usual "I need a quick favor." I could see Chris rolling his eyes on the other end of the line, but he helpfully obliged. In minutes he was in his geologic databases and pulling up information. The site had once been a volcano. It blew itself to smithereens in an eruption 3,700 years ago and left behind a crater that had filled with water. The volcano was still active.[4]

As Chris browsed the material, I moaned about how the annoying Italians had blocked off the caves and complained that I was an idiot for not thinking to have Charlie bring over his lockpicks. As I told him that I wanted to return and try getting into the oracle's bath cave at another time, he suddenly interrupted me: "Mate, that would be a really bad idea. It says here there's a serious danger of poison gas in that cave."

Lake Avernus has a long history of belching out toxic gases.[5] So much would be released that, periodically, all the birds in the area would simply drop dead as if struck by a death spell. They were being poisoned, but the Greeks didn't know that. The name Avernus comes from *aornos*, which means "birdless" in Greek. In October of 2005, locals found thousands of dead fish floating eerily on the surface of the lake. A geology team published a report in 2008 in the *Journal of Volcanology and Geothermal Research* that said this was the result of toxic hydrogen sulfide gas, which had collected at the bottom of the lake, suddenly launching toward the surface and killing all of the animals above.[6] It all seemed tantalizing to consider that our ancestors viewed the place as strong with dark magic, suspected that Hera, the goddess of shades, was active there, and decided it made a logical place for an oracle to take up shop.

Then I remembered why the name Avernus had seemed familiar to me earlier in the day. It was mentioned in the sixth book of Virgil's *Aeneid*:

> *Deep was the cave; and, downward as it went*
> *From the wide mouth, a rocky rough descent;*
> *And here th' access a gloomy grove defends,*
> *And there th' unnavigable lake extends,*

*O'er whose unhappy waters, void of light,*

*No bird presumes to steer his airy flight;*

*Such deadly stenches from the depths arise,*

*And steaming sulphur, that infects the skies.*

*From hence the Grecian bards their legends make,*

*And give the name Avernus to the lake.*[7]

Later that evening, while sitting at the bar in my hotel, one of the off-duty staff told me that, like the Greeks, the Italians hadn't fully appreciated the dangers of Lake Avernus either. He explained that they once gave tours of the oracle's historic cave until one day when the volcano released a puff of gas and killed a bunch of people. Thus the metal gates. I have not been able to determine if this is actually true. You'd think that something like "tourists killed by poison gas while visiting oracle of the dead" would make some headlines, but I've found nothing, so I'm skeptical about the tale. The official line from the Italian government? They claim a combination of Mafia activity and drug use in the caves led to their closure. None of this changed anything. I still yearned to get inside. I just had to make sure to bring along some supplemental oxygen, a bulletproof vest, and Charlie armed with a set of lockpicks.

## PASSAGES INTO DARKNESS

Similar to the oracle of the dead at Lake Avernus is a cave along a ridge in modern Turkey. Near an ancient Greek settlement called Hierapolis, the cave is wide enough for a person to enter, and quite deep too. Inside is a twenty-seven-foot-diameter grotto that was once surrounded by a handrail.[8] This hole caught the attention of many sadistic ancient tourists. Why do I say sadistic? Well, people went there to put animals inside and watch them die. This was often done as an act of animal sacrifice to the underworld gods, but some people tossed birds in for fun. The famed Greek geographer Strabo was one such tourist, who reported:

*Large enough to admit a man, but it reaches a considerable depth, and it is*
*enclosed by a quadrilateral handrail, about half a plethrum in circumference,*
*and this space is full of a vapour so misty and dense that one can scarcely see*
*the ground. Now to those who approach the handrail anywhere round the*
*enclosure the air is harmless, since the outside is free from that vapor in calm*
*weather, but any animal that passes inside meets instant death. At any rate,*
*bulls that are led into it fall and are dragged out dead; and I threw in spar-*
*rows and they immediately breathed their last and fell.*[9]

The effects on humans were the same. Even those who tried to brave
the depths of the cave by wrapping scarves or towels around their faces
rarely made it out alive.[10]

Unlike at the oracle of Delphi, where geology's role in making the
place what it was remained all but undetectable for centuries, at the
Hierapolis cave the dangers were obvious. Poisonous gases rise from
mineral hot springs.[11] Our ancestors came to the correct conclusion
that the gases killed. They then provided the only explanation that they
could by saying the cave was an entrance to the underworld. It came to
be known as the Plutonium or Pluto's Gate.

According to archaeologists, the lethal gas in the cave is carbon dioxide.[12] I knew that carbon dioxide could kill you at a high enough concentration, but I didn't think it was the sort of thing that would make you
drop dead instantly. That was more the territory of such stuff as cyanide
gas, which can knock you out in seconds. Yet, there are no reports of anything remotely similar to cyanide gas at Hierapolis. How could this be?

The first clue lies in the animals that were most often being sub-
jected to the dangers of the cave. Birds, particularly small birds such as
the sparrows that Strabo tossed in, are very vulnerable to toxic gases.
This is why canaries were so valuable to coal miners during the days
before electronic gas detectors. If the canary fell dead off its perch, the
miners knew they had to evacuate in a hurry even if they didn't feel ill.

I initially suspected our ancestors saw the birds respond in this way and assumed all other living things died just as quickly. As it turns out, the carbon dioxide levels in the cave were probably exceptionally high.

If we are exposed to air composed of 7 percent carbon dioxide for around five minutes, we start to suffer from confusion and feel out of breath.[13] At around 10 percent, we start to asphyxiate. At 20 percent, we pass out in a few minutes and then die.[14] At concentrations higher than that, the effects are faster. So what were the levels at the Plutonium? Try 91 percent.[15] "At ninety-one percent, you'd drop pretty fast," said toxicologist Alan Woolf at Harvard Medical School.

Among the stranger happenings at Hierapolis were the activities of the eunuch priests who worked at the site. Known as Galli, these men worshipped the local mother goddess, Cybele, who could purportedly fend off the site's deathly magic.[16] Strabo speculated that something about the eunuch's maimings granted them protection from the dangers of the site, but this seems unlikely.[17] While lopping off the testicles has a large number of biological effects,* there is no evidence that it would provide any protection against toxic fumes.

I suspected the eunuchs were holding their breath. Strabo himself suggests this by saying, "I could see in their countenances an indication of a kind of suffocating attack, as it were."[18] As to why something so mundane would appear supernatural to visitors, it probably was due to a mix of knowledge and practice. Most people assume that the amount of time that they can hold their breath is not something that they can increase, but this is untrue. The skill requires both physical ability and mental control. Do it a lot and you will get better. I also suspected that the Galli were amplifying their abilities by cheating. Since the Galli knew the site better than anyone else, I wondered if they had found sections of it where the

---

* With shutting down the sex drive and providing a lovely singing voice being just a couple of examples.

air was less foul and taken additional breaths there. While discussing this with Francesco D'Andria, one of the archaeologists responsible for excavating the Plutonium, my suspicions proved correct, as he explained that the Galli are noted in several ancient sources as having stood in slightly elevated places while in the grotto. When I double-checked this with one of my chemistry friends, he laughed and asked, "You got time for a joke?"

I nodded in confusion.

"Okay, so there was this canoe in a lake with a priest, a rabbi, and a Buddhist monk. The priest says, 'You guys wanna see a demonstration of my faith?'—and then steps out of the canoe and walks on the water over to the shore and back again. Amazed, the monk and rabbi both then try this, but quickly splash into the water and sputter as they struggle back aboard. The priest then looks at them and says, 'What? You guys don't know where the rocks are?'"

Carbon dioxide is 1.5 times heavier than air, so it sinks. When ejected out of volcanoes, the gas runs like an invisible river of death down canyons, suffocating all animals in its path. When it reaches ditches and valleys, it pools in them, creating vaporous lakes that kill anything present.[19] As year-round residents at the site, the Galli would know precisely how high the toxic pool of gas in the grotto was and perform their sacred rites standing on rocks that kept them just inches away from instant death. They may also have known something even more fascinating: that the underworld's power was weakened during the day.

Sun gods and underworld deities are almost always enemies in mythology. Aside from making logical sense, since light and darkness are antithetical, chemistry is behind this too. Carbon dioxide molecules readily absorb the infrared radiation found in sunlight. When this happens, the molecules vibrate and give off heat, and if the carbon dioxide gas has pooled on the ground, these vibrations dissipate it.[20] This would make the entry to the Plutonium more dangerous in darkness, and this is exactly what locals report today. Mice, cats, weasels, and foxes are much more commonly killed at the site on cloudy days and at night than when the sun is shining brightly. We now understand this as the result of chemistry, but it is easy to see how our ancestors viewed this as the noble sun god, Apollo, using his power to keep the shadowy Hades from expanding his reach from beyond the grave. Yet, not all passages into the underworld were associated with poisonous gas. Some had venomous guardians.

## GUARDIANS OF THE DEAD

Scattered throughout Mexico are stunningly beautiful pools of water. In a land scorched by the sun, the freshwater is nothing short of a wondrous gift to the thirsty traveler. To the Maya who lived there long ago, the pools were sacred.

They called these pools cenotes, and while the Maya did often use them to supply their cities with water, they also believed that they were gateways to the underworld.[21] During the height of the Mayan empire, people would travel for hundreds of miles to visit these pools and throw into their depths offerings to the gods of gold, jade, and pottery.[22] Humans too were sometimes thrown in, particularly during times of famine, prolonged drought, or disease. Individuals, usually captives, would be tied to posts, speared, and then hurled to their deaths to appease the dark deities below.[23] What is unclear is whether the Maya understood just how special their pools actually were.[24]

Divers exploring the cenotes have discovered much more than just the treasures that the Maya threw in.[25] They have found that the pools are not really pools at all but tunnels that travel for miles through limestone caverns and out to sea. Moreover, not all the water in them is fresh. All cenotes have a point where freshwater, accumulated from rain, meets salt water. These transition points, called haloclines, obscure vision and shimmer ominously. They do look like magical portals, and for a number of divers who have passed through them, they have literally granted a one-way trip to the land of the dead.

At the best of times, cave diving is dangerous. If something goes wrong, you can't just swim to the surface. Unsurprisingly, a lot of divers die in caves due to that fact. But many of us in the diving community have been vexed by the number of top-notch divers who have perished in these places.

Biologists Björn von Reumont and Ronald Jenner at the Natural History Museum in London are among the few intrepid scientists willing to study these dangerous cenote ecosystems. Working with expert divers, they went deep into the caves to collect the life there and came upon an unusual critter.

Distantly related to crabs and lobsters, the animal belonged to a group of crustaceans first discovered in 1981 called remipedes.[26] Pale, blind, and tiny, these creatures make their homes in underwater cave networks. Some, such as the one Dr. von Reumont and Dr. Jenner found, have needlelike structures growing out of their front claws.[27] They also spotted the cenote-dwelling remipedes tossing aside empty exoskeletons of shrimp.[28] This led them to suspect that these remipedes were using their needles to hunt prey.

Close examination of the needles in 2007 revealed hollow channels.[29] These were precisely the sorts of channels that would be good for injecting a fluid into whatever was being struck. Even more suspicious were the presence of dense muscles around reservoirs at the bases of the needles that seemed ideal for pushing liquid out into the channels. The fluid is a venom, but not just any venom. It is something of a mix between the venom found in vipers, which breaks down tissues, and the venom found in spiders, which attacks the nervous system and causes paralysis.[30]

While the remipedes have not yet been seen actively hunting, Dr. Jenner and Dr. von Reumont suspect they stab their prey with their needles, and then the toxin that attacks the nervous system sends their prey into uncontrollable spasms so it can't get away.[31] The tissue-destroying toxin next liquefies the shrimp inside its own exoskeleton so the remipede can casually slurp it up later.[32] That alone is pretty cool, but it's remarkable that, of the seventy thousand known crustacean species on our planet, these cenote-dwelling remipedes are the only venomous ones to have ever been found.[33]

What effect the venom has on people remains unknown, but being a diver myself and rather curious if the remipedes could be responsible for the inexplicable deaths of first-class cave divers, I asked Dr. Jenner about it.

"I've been wondering the same thing ever since we first saw those needles," he explained. "I mean, we so often hear about cave divers dying there under mysterious conditions. It makes me wonder if these remipedes just might be responsible."

If the remipedes are behind all of these deaths, then the Maya may have been more correct about their cenotes being passages to the underworld then they could ever have realized.

## TOUCHING THE VOID

Did the Maya know about the remipedes? I doubt it. The crustaceans live far too deep inside the caves for any human to encounter without diving equipment. That is not so much the case for the entrances to the cenote tunnels themselves, though. It would not have been difficult for the Maya to swim down and discover the dark caves. The same is true for the shimmering haloclines. While there are no records of this, I suspect the early Maya did know their sacred pools contained shimmering portals and ominous tunnels, and that this knowledge inspired them to come up with the "passages to the underworld" idea that is in their mythology.

The Maya did not just use their pools as places of sacrifice to appease their gods. On some occasions, unwounded children were thrown into the pools at dawn and then left to literally sink or swim for hours.[34] If the kids survived until midday, it was seen as divine intervention and they were hauled out with ropes.[35] Once breathing easily again, they were interrogated by the priests to find out what the gods had planned for the Maya during the coming year.[36]

Intriguingly, the Vikings may have engaged in a somewhat related practice. In "Hávamál," a poem from the Icelandic Codex Regius that was written down in the 1200s, Odin, the god of war and wisdom, claims that he can make men hanging from nooses speak with him.[37] This is fascinating, because the ritual hangings made by the Vikings were unlikely to have been neck-snapping affairs in which victims were dropped from a height and killed quickly.[38] The evidence that we have suggests that they were hoisted up just off the ground and slowly asphyxiated.[39] So common was this practice that one of Odin's alternate names was Hangadróttin, meaning "Lord of the Hanged."[40]

While this is terrible stuff, a lot of evidence suggests that the Maya and the Vikings weren't totally crazy in believing their gods were speaking to those struggling to survive as they ran out of oxygen. People who have had near-death experiences commonly say they saw bright lights at the end of a dark tunnel, felt the euphoria of having briefly stepped into heaven, or had conversations with angels. It all sounds like fiction, but it has a basis in reality.

Seeing light at the end of a tunnel happens naturally. When blood and oxygen supplies run low, the cells that process our central vision and the cells that process our peripheral vision respond differently, and we end up seeing a central bright light with a dark periphery.[41] Fighter pilots who fly at high speeds experience the effect for about five seconds as they accelerate and temporarily deplete the oxygen supply to their eyes. It is also found in glaucoma sufferers for much the same reason.[42]

The bliss of having visited the blessed fields of Elysium for a few minutes is not much different. The body can produce compounds that drive away pain and grant feelings of euphoria through opioid peptides and opioid receptors in the brain.[43] We know the opioid system often activates when animals are experiencing traumatic circumstances, and we suspect this happens to numb pain that might otherwise hinder the animal from getting away. As an example, it isn't helpful for a gazelle being attacked by a predatory cat to fixate on pain from a slash wound. If the gazelle is to survive, it is better for it to ignore the pain and focus on escape. This is a major reason why we think natural selection has given most species in the animal kingdom the ability to experience a euphoric high without the use of external drugs. With near death, the theory is that the opioid system sometimes kicks into gear even when its activation might not make much of a difference. Remember, the bio-chemical pathways of our bodies are not entirely logical. If evolution has driven the opioid system to automatically start when chemicals associated with acute stress are detected in the bloodstream, the system won't necessarily distinguish between the stress associated with being mauled by a bear and the stress experienced when suffering a heart attack. Thus the euphoria and bliss that people report from near death is not likely the result of their having briefly stumbled into heaven, but is a natural high that has been used by the body for millennia.[44]

As for seeing dead people or chatting with angels, patients with Alzheimer's and progressive Parkinson's disease often have these sorts of experiences. The same is true for people suffering from extreme exhaustion or with advanced macular degeneration. A 2005 paper in the journal *Neuroscientist* revealed that electrically stimulating the region adjacent to the angular gyrus in the brain, which is associated with spatial cognition, makes people feel as if someone were standing nearby even if they are seated in an empty room.[45] Moreover, an experiment led by Jimo Borjigin at the University of Michigan and reported in *Proceedings of the National Academy of Sciences* in 2014 found that rats induced to have heart attacks showed a surge of gamma oscillations during the thirty seconds that followed the attack.[46] These oscillations indicated that their brains were highly aroused and processing a whole lot of information as the rat grim reaper* was coming to claim them. Thus, a brain under assault appears to

---

* *SQUEAK.*

have a number of ways to make a person believe he or she is being visited by a long-dead relative, an angel, or even a Mayan god. There is nothing paranormal about the experience, but it is real nonetheless.

What remains mysterious is why these experiences appear in some people but not in others. In one study, published in the *Lancet* in 2001, a team led by cardiologist Pim van Lommel at the Netherlands' Rijnstate Hospital, extensively interviewed 344 heart-attack patients within a few days after cardiopulmonary resuscitation (CPR) was used to save their lives.[47] All of these participants had been nonbreathing and without a pulse (i.e., dead) for a time before CPR was initiated, yet Dr. van Lommel reported that only 18 percent of the patients said they had a near-death experience.[48] These 18 percent reported classic stuff such as having an awareness of being dead, observing a celestial landscape, meeting with deceased people, and seeing lights and colors. Dr. van Lommel and his colleagues studied this pool of patients to see if they could work out what, if anything, led this fraction of their study group to have these experiences. The team looked at age, education, religion, feelings of fear, foreknowledge of near-death experiences, medications, and the details of each patient's medical circumstances.[49] The key factor was age.[50] Patients younger than sixty were significantly more likely to report having a near-death experience than those over sixty. Why this is remains to be determined, but Dr. van Lommel postulates that it is probably because our memories work less well as we get older.[51] What was truly extraordinary was that, when the researchers went back to interview patients two and eight years later, the people who had had near-death experiences still remembered them with exceptional detail years later.[52]

Was interrogating people who had near-death experiences in cenotes or who had been slowly asphyxiated at the end of a rope on a tree a viable method of predicting the future in the way that ancient Greeks looked at bird migrations and animal livers? I find this unlikely. While I can see how the Maya came to believe that their gods spoke to those who nearly died in their sacred pools and how the Vikings believed Odin spoke to men gasping for air as they hung from trees, it feels like too much of a leap to think that the hallucinations of these victims changed based upon larger climatic conditions in the area.* However, what I don't think is much of a leap is to use these research findings in the realm of near death to argue that Odin's title "Lord of the Hanged" really had very little to do with his war portfolio and much more to do with his role as a god of wisdom and prophecy.

A near-death experience is not the only thing that can mess with our minds, though. Numerous other experiences that don't bring us to the brink of death are perfectly capable of changing how we perceive the world. In that vein, let us explore enchantments.

---

* Sure, in the case of the Maya it is possible that water temperature and salinity in the cenotes might shift with El Niño and La Niña conditions, and this might lead survivors to have different hallucinatory experiences, but there isn't any literature on this. Seriously, can you imagine what a proposal for an experiment of this sort would look like? "Research team seeking young children to be flung into pools of varying temperatures and salinities for several hours at a time. Survivors to be interviewed."

# ENCHANTMENT

## Love Potions, Sleeping Drafts, Mind-Altering Fruit, and Penises Pierced with Stingray Spines

*Dip the apple in the brew. Let the sleeping death seep through.*

—THE QUEEN, *SNOW WHITE AND THE SEVEN DWARVES*

With their ability to ensnare the heart and alter the mind, objects of enchantment have fascinated civilizations from the beginning.

Scholars fiercely debate when the Torah was first written down. One portion of the academic community argues it was written around 1400 BC, but others suggest that the text was first written between 800 and 400 BC. Regardless of who is right, there is little question that the book of Genesis contains one of the earliest and most famous forms of enchantment magic: the apple of Eden.

Essentially given free roam of the Garden of Eden, Adam and Eve lived in paradise. They ruled over all other animals, could do pretty much whatever they pleased, and suffered no disease or hardship. The land provided for them and life was good. There was only one rule. God commanded, "You may freely eat of every tree of the garden, but of the tree of the knowledge of good and evil, you shall not eat. For in the day that you eat of it, you shall surely die."[1]

Adam and Eve initially heed this warning and stay away from the tree. However, temptation eventually gets the better of them. Eve meets a serpent in the garden and is told that God has been lying. According to the serpent, the fruit from the tree is beneficial: "You will not die. . . . When you eat of it, your eyes will be opened, and you will be like God, knowing good and evil."[2]

Led by the serpent's words, Eve plucks a piece of fruit* from the tree, takes a bite, and then shares it with Adam. The effects of the fruit are swift: "Then the eyes of both were opened, and they knew that they were naked; and they sewed fig leaves together and made loincloths for themselves."[3]

While the fruit from the tree of knowledge is never specifically described as enchanted, the implication is obvious. The Garden of Eden is a divinely crafted place, God wanders its groves, nutritious food is always available in abundance, foul weather is absent, and the animals can speak.** There is also the effect that the fruit from the tree has: Adam and Eve immediately became aware of their nakedness and covered themselves.

The fruit from the tree of knowledge is mind-altering. Adam and Eve came to see themselves and the world around them differently from how they had before. Their minds were opened and forever changed.

It is easy to dismiss the story in Genesis as fiction. I certainly did for the better part of my life, but then, as I was working on an article about several species of parasites that commandeer the minds of the animals that they infect, I got to wondering whether there might be something real behind the tree of knowledge.

## MAGIC MUSHROOMS

Some compounds in the natural world can alter human awareness. The effects are similar to those that the fruit in Eden had, but they are not found in trees. They are found in fungi.

---

* Art usually shows this as an apple, but the Hebrew text only says "fruit."
** Well, the snakes do, at any rate.

Many species of mushrooms are toxic. Eat them and you *will* surely die. This might be bad for you, but from a fungal perspective the situation works out pretty well, since being toxic keeps other organisms from moving in and consuming food that the fungus is already breaking down. Fungi are quite literally killing off competition. Even so, being toxic is not the only survival strategy that has evolved among fungi.

Some fungi just make animals that feed upon them ill. This is not generally advantageous to fungi fed upon by dumb animals. But intelligent animals that can learn from their experiences will choose to never again eat the fungus. Things get even better for the fungus if the intelligent animals are social, because after one animal is made ill, others in the community, particularly youngsters, will learn to avoid that fungus as well. Under such a scenario only one fungus-infested food source needs to be consumed for an entire population of potential consumers to be put on alert. In contrast, if a social animal were killed outright, it would have no ability to teach others, and many more animals would come—a situation that is bad for the fungi and worse for the animals.

Most fungi that have hallucinogenic effects are not much different from those that cause illness. These fungi, popularly known by communities where tie-dyed shirts are popular as "magic mushrooms," contain compounds that mess with the minds of the animals that eat them. Exactly what other species experience is difficult to determine, since it is hard to understand how a deer or an elk sees the world when tripping out, but in humans these hallucinations vary with the state of mind. When people are tense, nervous, or frightened, hallucinogenic mushrooms can be the stuff of unspeakable nightmares. We think experiences are similar for most wild animals.

For animals, these hallucinations likely drive them to henceforth avoid consuming fungus-infested food. "Once horses graze on grass tainted with these fungi, they will never touch those grasses again,"* fungal biologist Keith Clay at Indiana University explained. This suits the mushrooms just fine, since they benefit from not having their food eaten by others.

Our earliest hunter-gatherer ancestors probably responded in the same way. They ate, they had a bad trip, they ran, and they never came back. However, as we evolved, so too did our interactions with hallucinogens. Through trial, error, careful control of the amounts consumed, and social learning, people realized these mushrooms could induce powerful experiences. Today, biochemists studying the compounds in these fungi know that if people consume them under stressful circumstances, they are likely to become terrified; but if they consume them in environments where they feel safe, the effects are usually positive.

Eager to know more, I turned to medical biochemist David Nichols at Purdue University. "When people take psilocybin, the compound in 'magic mushrooms,' in a calm place, it is common for them to feel spiritual and experience something akin to a brain reboot," Dr. Nichols explained. This left me fascinated and uncertain. *Brain reboot* wasn't exactly terminology that I'd encountered before.

As I sat there in a long pause, Dr. Nichols advised that neither he nor anyone else was certain of what the brain was doing when exposed to psilocybin: "There is a fundamental shift in the functioning of the brain. The cortex starts processing information like crazy, but we don't think external stimuli are being processed, because you can poke and shout at people in this state and they don't respond. They appear to be in another reality."

---

* Some species are exceptions, such as reindeer, which go out of their way to eat mind-altering fungi. The reindeer help the fungi to spread their spores by defecating them in far-off locations. We think this is a bit similar to the way in which plants give honeybees nectar, in return for which the insects carry off pollen to other plants, except that the reindeer are rewarded with both food and a high.

This effect of increased processing with a disconnect from external stimuli is leading Dr. Nichols and his colleagues to speculate that the brain is being drawn into interacting with primordial regions of itself that contain deep-seated emotions, memories, and feelings. "It is conceivable that these drugs are forcing the cortex to work with brain regions that existed before humans were even humans," he said.

Neurologists and biochemists such as Dr. Nichols suspect that this forced interaction between the brain and its primitive parts might be a lot like a near-death experience, where awareness is dramatically changed. However, instead of the drugs bringing people to near death, they are bringing them to a state that existed prior to the evolutionary birth of the human brain. It is wild stuff, but the work has a practical side, because the fungal compounds are showing a lot of promise in medicine.

For patients who cannot kick addictive habits, psilocybin can be an invaluable lifeline. In one recent study, published in the *Journal of Psychopharmacology*, 80 percent of a group of smokers who had tried to kick their habit multiple times over the years finally quit after being treated with psilocybin.*[4] Crucially, they remained off cigarettes during the following six months. Psilocybin is also proving useful in people with terminal diseases who have fallen into severe depression. One research group led by Charles Grob, a psychiatrist at the University of California, Los Angeles, has run experiments with these sorts of patients and published the results in the *Archives of General Psychiatry*. Psilocybin helped patients snap out of their depression.[5] "Almost all of our patients commented that the drug gave them a new sense of perspective on their lives and that they could finally see both themselves and the world with new vision," commented Dr. Grob. Said another way, it was as if their eyes were opened.

---

* This is staggering when you consider that nicotine replacement therapy, the current strategy for helping folks stop smoking, is only effective 7 percent of the time.

While extensive research on the effects of these fungal compounds on the human brain is a recent phenomenon, deliberate use of them is not. We know mushrooms containing psilocybin have been used intentionally for thousands of years. Could this mean that the "fruit" on the tree of knowledge was grounded in something authentic? I pulled up one of the oldest versions of the Torah that I could find at the British Library, but as my Hebrew isn't great and I am hardly a scholar of Jewish literature, I was going to need some help; so I called the most learned rabbi I knew.

When the words "So might Adam and Eve have been taking psilocybin?" came out of my mouth, I was immediately glad that I was speaking with a rabbi who was more than just a professional contact. As a kid I went skiing with Rabbi Eli Herscher in the mountains of California, and as a teen I drove his son Adam to school for a couple of years. Heck, he played a major role in my wedding with Thalia, helping us to weave in Jewish cultural elements alongside Greek and Finnish ones. I could speak my mind to him without any concern of offending. Thank goodness.

"Well, there are some pretty wild visions in the Torah," said Rabbi Herscher slowly and matter-of-factly, after taking a few moments to contemplate the subject. "I mean, there's this one section where the prophet Ezekiel has some very powerful visions of God's chariot with wheels of fire and diamonds, and you really wonder: What exactly was he on? Yet I don't know of any evidence that what Adam and Eve ate from the tree was anything more than a fruit. But I'll be honest with you, Matt, just because I don't know of anything doesn't mean a lot."

Our conversation descended into contexts and definitions, and we eventually focused on the single word in the Torah representing what Adam and Eve plucked from the tree and ate: פְּרִי. This sounds a bit like "pree" in English when the proper vowels are applied, and modern Hebrew defines the word as "fruit." However, as we spoke, Rabbi Herscher pulled out what was effectively the Hebrew version of the *Oxford English Dictionary* and started making all the sorts of noises that one expects to hear from a scholar pondering intriguing ideas.

"Huh, well, פְּרִי actually turns up in a number of historic constructs. It is part of פרי העץ, which literally means 'fruit of the tree.' It is also part of פרי האדמה, which means 'fruit of the earth.'" Fascinating. What precisely did our ancestors mean by this? The dictionary stated "vegetables of every different kind including grasses." Were mushrooms vegetables by this definition? "It really seems open to interpretation," advised Rabbi Herscher.

Does that mean some sort of reality lies behind the Genesis tale? It depends upon how you look at it. I sincerely think that Adam, Eve, and the Garden of Eden were fictional, but I also suspect that the author of the story knew the natural world had objects capable of opening eyes and minds. The author might even have tried some.

It is not surprising that reality is found creeping into Genesis because, as odd as the idea of God's creating humanity in a garden might sound to a nonbeliever, the story is loaded with explanations for why the natural world works in the ways that it does. Eve is created from Adam's rib, which for ancient people proved a helpful explanation for why women and men share similar skeletal structures but differ so significantly in physical appearance.[6] For picking the fruit from the tree of knowledge, Eve is cursed by God to experience terrible pain when giving birth, which helped explain to ancient believers why the sacred activity of being fruitful and multiplying hurt so much and killed a lot of women during delivery.[7] The serpent is also cursed by having its legs removed and being forced to forever slither on its belly in the dirt, providing an explanation for why snakes have no legs.[8]

All of these explanations in Genesis strengthened the appeal of the work to those who listened to the story because they helped make sense of an otherwise confusing world. I suspect the fruit from the tree of knowledge was a part of this explanatory process. Nobody understood why the people who consumed certain fungi were having eye-opening and permanently life-changing experiences. By suggesting that these mushrooms were relatives of the "fruit" from Eden, Genesis was providing a logical—and valuable—explanation. Of course, Western civilizations were not the only ones enchanted by such things.

## BLOOD MAGIC

In the *Popol Vuh*, a compilation of epics that the Maya passed from generation to generation and ultimately allowed outsiders to write down in AD 1700, the creation of the world is explained as happening not once but four times. During the first three attempts, the gods tried, and failed, to create a world with beings that were carved in their own image.[9] These three first attempts weren't all bad. The gods saw steady improvements but were unhappy with the results and started over each time. Finally, in their fourth attempt, they got things right. Well, actually, they got things too right.

The *Popol Vuh*, which means "Book of the People," says that the humans created during this fourth trial "saw everything under the sky perfectly . . . understood everything perfectly."[10] The gods had created perfection and viewed this as bad. They became afraid of the people they had labored to produce and attacked humanity to protect themselves. The first man was struck in the eyes by a divine bolt so that his "sight" was injured and he could no longer "see" with the godlike understanding that he had been created with.[11]

Sound eerily familiar? Why are the gods in Mayan mythology so desperate to keep humanity from having godlike understanding of the world around them, just as God in the Judeo-Christian Bible is so keen to protect Adam and Eve from the tree of knowledge? Theologians and historians have debated this fascinating question for decades. Did a single story about humanity being created without knowledge of the world emerge long ago and simply end up in two different forms? Did Judeo-Christian mythology brought over by early missionaries contaminate the *Popol Vuh* before it was written down? Nobody knows for sure. However, we do know that reclaiming godlike vision was an essential aspect of Mayan culture.[12]

As part of attaining godly sight, mushrooms were consumed. But during their most sacred rituals, the Maya drank a fermented beverage—made from the bark of the balché tree—that produces moderate intoxication.[13] They also ate mildly poisonous morning glory seeds and toxins from the peyote cactus to push them into an even more altered state of consciousness.[14]

After dining on their hallucinogenic smorgasbord, the Maya performed rituals that sound nasty from a modern perspective. Women would pierce their tongues with richly ornamented flint spikes, shards of volcanic glass, or stingray spines.[15] They would then draw a string through the resulting bloody hole. If that weren't bad enough, thorns were often woven into these strings. The pain was probably intense.* Rather than pierce their tongues, the Mayan men pierced their penises.[16] Yeah, you read that right. They'd poke a hole in their sexual organs and then run a string covered with thorns through it.

For years we've had to speculate on how such painful activities messed with the human mind, because scientists are, for better or for worse, simply not allowed to eviscerate study participants. Even so, keen researchers have found ways around this problem by devising methods that cause considerable pain to study participants that ethics boards have been willing to approve. Enter the cold press.

You know how it really hurts to put your arm in water that is just

---

* I can only guess, as no literature exists on the degree of pain generated by drawing thorn-laced string through a pierced tongue, and no, thank you, I'm not testing this one out.

a touch above freezing? I know it all too well after sticking my arm in the freezing-cold Finnish ocean that Thalia leaps into after sitting in her uncle's sauna for a half hour every April. The lovely thing about this sort of pain is that, while it really hurts, it doesn't cause lasting damage. So research teams have explored the effects of stress and pain by having people stick their arms in cylinders of near-freezing water for stretches of time that are long enough to hurt but brief enough to avoid causing any legal entanglements.

I first encountered the cold press a number of years ago when writing an article on how stress induces stock-market traders to make illogical economic decisions. But it has since cropped up in a whole bunch of interesting studies. The most intriguing, and relevant to the Maya, appeared in January 2014 in the journal *Appetite*.[17]

In that study, led by psychologist Brock Bastian at the University of Queensland in Australia, thirty-six people were asked to insert their hands into a container filled with water. Inside the water was another small container with a hole in it, and next to it were a number of loose metal balls. Participants were asked to pick up the metal balls and place them, one at a time, into the container. The catch? Half of the participants were doing this in room-temperature water and the other half were doing it in water that was only one degree above freezing. After this, all participants were asked to rate how painful the experience was on a 0-to-5 scale where 0 was "no hurt" and 5 was "friggin' agonizing!"* The researchers then invited everyone to participate in a separate activity described as a "consumer study." Of course, it was no such thing.

Curious about whether pain would alter perceptions of flavors in food, Dr. Bastian and his colleagues asked participants to sample a chocolate cookie and rate, on a 7-point scale where 7 was "very enjoyable," how they

---

* Actually, it was "hurts worst," but that just doesn't seem to sum up the sentiment quite as well.

liked it.[18] Those who were exposed to the painful* cold press gave the cookie an average score of 6.03, while those who were picking up balls in room-temperature water gave the cookie a less enthusiastic 4.97.[19]

The findings left Dr. Bastian with a question: Was pain simply enhancing the positive experience of eating the cookie, or was it somehow increasing all of the body's awareness? With this in mind, a slightly altered experiment with less pleasant tastes seemed appropriate.

Using sugar, lemon juice, salt, and the bitter substance quinine, Dr. Bastian and his colleagues created four different solutions. Each represented key flavors detectable to the human palate—sweet, sour, salty, and bitter—and were presented to participants either immediately after experiencing the cold press or ten minutes after experiencing it. As with the first experiment, the participants were asked to rate the intensity of the tastes that they encountered on a 7-point scale where 1 was "not at all" and 7 was "very much so."

Those who tasted the solutions immediately after the pain experience gave the sweet solution an average score of 4.15, the sour solution a 4.09, the salty solution a 4.65, and the bitter one a 4.94.[20] In contrast, participants who waited ten minutes before taking the taste test gave these flavors scores of 3.35, 3.59, 3.65, and 4.24, respectively.[21] Pain was clearly enhancing the perceived intensity of all the flavors, not just the pleasant ones.

In one final examination, Dr. Bastian reran the initial experiment, but rather than giving people cookies, he gave them ten cups of water. Some of these cups were just water, but some contained tiny amounts of unsweetened flavoring, such as ginger, peppermint, coconut, apple, and orange. While the participants were unaware of it, they were all given water for their first three cups and only given flavoring in their remain-

---

* Dr. Bastian had to remove one participant from the study because he kept his hand in the icy water for nearly seven minutes! Since this was more than three standard deviations above the one minute and thirty-six seconds that most people could tolerate, it suggested that this participant's experience of the cold was very different from everyone else's. This doesn't mean that such extreme tenacity isn't interesting—it very much is— but we will get to this and other superhuman feats in the next chapter.

ing cups. More important, the dose of this flavoring was not random. It steadily increased by half a milliliter with each successive cup from the fourth to the tenth.

Once again, pain seemed to grant heightened sensitivity. Those who had stuck their arms in the cold press tended to correctly identify flavors by cup 8.17 on average, while those who had stuck their arms in tepid water got things right by cup 9.83.[22]

I asked Dr. Bastian to explain why pain was influencing taste in this way, and he said, "We suspect pain is capturing attention and focusing it on bodily sensation. If you think about it, this is how pain fulfills its role as an alarm signal. It draws attention to something that is wrong with the body. When the pain stops, we remain in this vigilant state for a period, meaning we are more sensitive to other sensory experiences that follow the pain."

For years, scholars have argued that the bloody practices of the Maya were simply dramatic displays of faith.[23] It seemed logical to argue that jabbing a shard of obsidian or a stingray spine through a tongue or a penis was an impressive way for leaders to show that they were fully devoted.[24] The arguments all made sense, but, based upon Dr. Bastian's research and the material in the *Popol Vuh*, I wondered if there was more.

If pain grants an enhanced ability to detect flavors, would the Maya have concluded that these painful rituals helped them regain the divine sight that they once had? The *Popol Vuh* is unclear about what "seeing the world as gods" meant, but it seems reasonable that the Maya felt their gods could see more colors, hear more detailed melodies, feel more textures, and, yes, taste more in their food than mortals would.

While Dr. Bastian's work has, so far, only revealed that pain enhances the human ability to taste, his findings hint that experiments exploring the ways in which pain alters the senses of touch, sight, and sound may be well worthwhile. "These are questions that just haven't been answered yet," Dr. Bastian explained. I would not be surprised if, in time, we find that the painful rituals of the Maya allowed them to "see" in a profoundly different way.

## SWEET DREAMS

Closely related to the awareness-altering enchantments found in the fruit of Eden and the blood magic of the Maya are those that can cause one of the most powerful awareness alterations of all: sleep.

In days long gone, the requirement of sleep was frequently a threat. Sleep left our ancestors vulnerable to theft or predation.* Going to sleep has historically been scary. It is almost certainly why humans have an innate fear of the dark. For all of these reasons, anyone who could command sleep and force it upon others was viewed as supernaturally powerful.

In Greek mythology, Medea was the niece of the sorceress Circe, who herself was rumored to be the daughter of Hecate, the goddess of magic. While Circe could transform men into animals, Medea had her own gift. She could induce sleep.

Her tale, as told by the Greek scholar Apollodorus around AD 200, started with love. During his travels, the hero Jason, of the Argonauts, met Medea, and she immediately fell for him. When she learned Jason was searching for the Golden Fleece, which was said to be "guarded by a sleepless dragon," Medea quickly offered her magical services.[25] After a long journey, the two of them came to the grove where the fleece was found hanging on a tree, and Medea put the dragon guarding it to sleep with her drugs.[26]

Like the story of Adam and Eve, this is a fictional tale, but it too probably has kernels of truth. The Greek word that is so often translated as *dragon* is *dracos*, which can also mean *snake* or *serpent*. Indeed, in many translations of Jason and Medea's adventures a great snake guards the fleece

---

* I haven't had a single restful night in the tropics after writing about how effective large pythons are at sneaking into Indonesian huts and eating people in their sleep for the dragon chapter in *The Science of Monsters*.

rather than a dragon. Did great serpents once live in the Mediterranean? Yes. As for Medea's sleep-inducing drugs, these were probably real too.

A number of plants, when consumed, cause people to fall asleep quite suddenly. Three of the most potent species that have this effect—belladonna (*Atropa belladonna*), henbane (*Hyocyamus niger*), and Jimsonweed (*Datura stramonium*)—belong to the deadly nightshade family.[27] All three, as their family name implies, are deadly at high doses, or if the wrong bits of them are consumed. However, as I discussed in the chapter "Transformation," in small quantities and with the right bits administered they can have other effects. When properly prepared, all have the potential to relieve pain, dramatically slow down breathing, and reduce heart rate without killing.[28] The effect is not really sleep, but would have looked a lot like it.

"In British folk medicine alone we have seen henbane, woody nightshade, and lesser periwinkle used quite a lot as sedatives and soporifics. So the idea of sleeping potions having been brewed from them is perfectly possible," explained botanist William Milliken at the Royal Botanic Gardens, Kew. And belladonna, henbane, and thorn apples would all have been available during the days of the ancient Greeks.[29]

While Medea is described as a witch, she is always using ointments, potions, and drugs to do her work. Her power came from a knowledge of botany.

Sleeping potions play pivotal roles in medieval tales of the wizard Merlin, who creates an elixir to knock out the Duke of Cornwall so King Uther can secretly sleep with the duke's wife and conceive the child who grows up to be King Arthur.[30] These potions also appear in many of Shakespeare's plays, such as *Romeo and Juliet*, where Juliet drinks an alchemist's potion that makes her look dead but actually throws her into what the Bard calls "very deep sleep."[31] They also appear in *Cymbeline*, where a doctor is ordered by the wicked queen to create a poison for one of her enemies but disobeys and creates a powerful sleeping potion instead that makes the queen think her enemy is dead when he is not.[32] A sleeping potion is also found in the form of a magically poisoned apple in the tale "Snow White and the Seven Dwarfs."[33] Yet alongside the frequent use of sleeping potions rises a new form of enchantment, that of love magic.

## LOVESTRUCK

In ancient Greece the gods could make mortals fall in love with whomever or whatever they liked. There is no explanation for how this was done; the gods just willed it and it happened. Most people enchanted by the gods were driven to fall in love with people, or animals, that they would never normally have fallen for. For example, Aphrodite enchants King Minos's wife to fall in love with a bull and conceive the Minotaur as a punishment for her husband not making appropriate sacrifices.[34]

Our ancestors fell in love for seemingly inexplicable reasons just as we do today, yet it is less of a mystery now. We know that a combination of chemical and psychological cues kindle the flame of attraction. Nevertheless, even with this knowledge, the best psychologists and biochemists in the world still cannot reliably predict who will fall for whom. Thus it is understandable that people historically believed the gods were using their powers to fiddle with romance.

Around AD 160 a noted Roman author named Apuleius was accused of using love spells to make a wealthy widow fall in love with him.[35] A court case ensued, and Apuleius, being a masterful writer, argued that love was strictly controlled by the gods. The prosecution's case fell apart and Apuleius was vindicated.[36] Even so, the power of love did not remain in the hands of the gods for long.

By the Middle Ages, in the tale of *Tristan and Isolde* a potion is used to make one character fall in love with another.[37] In Shakespeare's *A Midsummer Night's Dream*, faeries of the forest pour the juice of a flower on peoples' eyelids as they are sleeping to make characters fall in love with one another.[38] Shakespeare doesn't just describe some generic flower; he is specific in much the same way that Homer is specific in describing moly in *The Odyssey*:

*Yet marked I where the bolt of Cupid fell:*
*It fell upon a little western flower,*
*Before, milk white, now purple with love's wound,*
*And maidens call it love in idleness.*[39]

So it was a flower called "love in idleness" that had once been white
and, according to myth, became purple after being shot by one of Cupid's
arrows. This flower is real. Its scientific name is *Viola tricolor*, but its common names range from heartsease and heart's delight to tickle-my-fancy
and Jack-jump-up-and-kiss-me, which is where the modern American
name Johnny-jump-up comes from.[40] More intriguing still is that this
flower species has long been used to treat a variety of conditions, from
respiratory problems to inflamed skin. Today, compounds from the
flower are found in medications for bronchitis, whooping cough, and
skin disorders such as eczema.[41] As for love, there is no evidence that the
flower has any real effect on the hearts and minds of people when sprin-
kled on their eyelids, and there is little to suggest our ancestors made use
of it in this way.[42] It was not the only useless love ingredient, though.

High on the ingredients list of most love potions were the ground-
up testicles of animals, because they represented sexual power.[43] Sap
collected from specific trees was also valued, because trees were viewed
as symbols of reproduction.[44] Some love recipes even called for a woman
to take off her clothes, smother herself in honey, and roll around in
wheat. Then she was supposed to scrape off all the wheat that had stuck
to her body, grind it up, knead it between her thighs, and bake it into a
loaf of bread to be fed to a man whom she wanted to have fall for her.[45]

Another recipe involved a fish. Step one, be female. Step two, place a
living fish inside your vagina and wait until it dies. Step three, take the
fish out and cook it. Step four, feed it to the man whom you want to love
you.[46] It all sounds horrid, but it has a crazy kind of logic. The female
womb was widely believed to be the source of love during the medi-

eval period, and if you could infuse food with this "essence of womb" and then feed it to people in a directed manner, you could control who developed amorous feelings for you. Other love potions were made from menstrual blood and female urine for similar reasons.[47]

Oysters too were considered magical, because if you stare at an oyster long enough,* it starts to look like the female genitalia.[48] Because of this resemblance, feeding raw oysters to people was said to unlock feelings of love. This has been the basis of the explanation for why oysters are associated with romance for decades, but recent scientific analysis of oysters is hinting there's more to it.

A team of researchers from the United States and Italy reported at the national conference for the American Chemical Society in San Diego in 2005 on the chemical composition of Mediterranean oyster tissues.[49] They contained both D-aspartic acid and N-methyl-D-aspartate. Both of these chemicals are known, from previous work conducted on laboratory animals, to increase the release of many hormones, including the sex hormones estrogen and testosterone. Thus, the team argued that oysters might enhance sexual feelings or even trigger arousal. Then a 2012 paper published in *Advances in Sexual Medicine* found that D-aspartate as a nutritional supplement both improved sperm mobility and increased sperm concentrations in seminal fluid.[50]

Oysters do not appear to have been alone. The medieval Arabians made use of *ambra grisea*, a brownish-gray compound found in the guts of sperm whales.[51] This smelly stuff contains tricyclic triterpene alcohol, which, when consumed, raises testosterone levels.[52] Similarly, *Eleutherococcus senticosus*, better known as Siberian ginseng, increases stamina and nerve stimulation across the body and makes sexual feelings more powerful.[53]

Science has come a long way since the days when oysters and ginseng were first being tinkered with, and we now have a number of other

---

* I suggest helping this process along with a few glasses of chardonnay.

drugs capable of meddling in arousal and reproduction. Viagra rates at the top of this list in most people's minds, but, like the compounds used hundreds of years ago, it merely increases sexual performance rather than building the bonds of love. However, a couple of drugs being widely studied today look like they may go a lot further.

One of these drugs is oxytocin, a hormone naturally produced by humans and other animals during erotic interactions. In recent experiments reported in the journal *Biological Psychiatry*, researchers found that when couples were given a nasal spray of either oxytocin or water before having a discussion that would inevitably lead to their having conflicting opinions, those dosed with oxytocin smiled more, threatened less, and had reduced levels of the stress hormone cortisol in their saliva afterward than couples who had been sprayed with water.[54]

The man-made chemical 3,4-methylenedioxy-N-methylamphetamine, or MDMA as it is widely known, has similar effects. After conducting extensive tests with it in 2010, researchers noted, again in *Biological Psychiatry*, that those exposed to the drug showed increased friendliness, playfulness, and warm feelings for others while also experiencing a reduced capacity to recognize facial expressions associated with fear or anger.[55] This combination of effects led individuals to be less socially inhibited than they might otherwise have been and created intimacy among people who would normally not be intimate at all.[56]

Could blasting two people with a combination of Viagra, oxytocin, and MDMA just before a first-time meeting increase their sexual interest in one another or even inspire a romantic relationship? I wasn't sure, so I sought out some help.

Eli Finkel, a social psychologist at Northwestern University, describes his areas of expertise as "relationships and motivation," but really he's all about love. He has spent years analyzing the initial romantic spark and why long-term love sometimes follows and sometimes does not.

"We've known since the 1970s that if you encounter people under pleasant circumstances, you have a much higher chance of liking them than if you encounter them under unpleasant circumstances," explained Dr. Finkel. However, when pressed on whether "liking" somebody translated into attraction, he became a bit more hesitant. "*I like* versus *I'm attracted* . . . I'd have to go with yes based upon our speed-dating work."

During speed dating, women are usually seated at tables around a room while men rotate, spending a few minutes with each woman before moving along. Afterward, participants select whom they want to see again, and if a person they selected chooses them too, contact details are provided to them. Dr. Finkel has effectively hijacked speed dating in the name of science by offering research speed-dating events where video cameras are set up all over the room, recording systems are placed on every table, and participants are given a discount for consenting to have their amorous activities captured.[57]

"The fact is that when we give our participants questionnaires at the end of their sessions, if they say they 'like' a person, they almost always say that they are 'attracted' to the person as well. When we run the numbers on this, the correlation is very high," Dr. Finkel said.

As for using chemistry to mess with love, Dr. Finkel leaned back in his chair and sighed. "I don't know. . . . It's hard to imagine lust without biochemistry. Impossible, really, because we know estrogens and testosterone play such important roles. I mean, seriously, amp somebody up on testosterone and they will definitely be lustier. It's seductive to think this is true, but then you get into matters of free will and the complexities of relationships, and I start to get cautious."

To understand whether a love potion could be created, it is critical to tease apart the different aspects of loving relations and identify the biochemistry underlying each of them. This area has received a lot of attention during the past fifteen years from biological anthropologist Helen Fisher at Rutgers University.

She pointed out in the *Journal of Sex Education and Therapy* in 2000 that romantic relations have three key aspects.[58] There's lust, which is characterized by a powerful desire for sexual gratification. There's attraction, which is characterized by focused attention on a potential mate and is tightly linked to feelings of exhilaration when near that person and a need to develop emotional union with them. Finally, there is attachment, which is the calm and close social contact that most of us think of when we consider what marriage is like after four or five years of living together. While we most often think of this attachment as the final love system that activates in the building of a relationship, it isn't always. Dr. Fisher points out that these are systems and not stages. Attachment can happen between two people first, then attraction and then lust, or vice versa. Attraction may even be followed by lust and then attachment. Crucially, Dr. Fisher also found that specific chemicals are associated with these different love systems.

Sex drive is, unsurprisingly, closely associated with estrogens and androgens, hormones—such as testosterone—that guide and control the activity of the sexual organs. Attraction, she found through analysis of the behaviors of people who were lovestruck, is associated with high levels of catecholamines.[59] These include dopamine, which plays a key role both in transmitting reward signals in the brain and in focusing attention on details, such as eye color, waist-to-hip ratio, and skin quality. These also include norepinephrine, which is connected to vigilant concentration and imprinting, in the way that a baby duckling imprints its mother when it first sees her.

Intriguingly, Dr. Fisher also came to suspect that people who were deep in the throes of attraction, or lovesick, also had low levels of serotonin.[60] In her interviews with the love obsessed, these poor souls reported that they spent more than 85 percent of their waking hours thinking about the person they were smitten with.[61] This qualified as obsessive behavior in Dr. Fisher's book and led her to look at treatments for patients suffering from obsessive-compulsive problems. The drugs of choice used by psychiatrists were serotonin reuptake inhibitors, which would make the neurotransmitter serotonin more readily available.[62]

As for the chemicals behind attachments, Dr. Fisher knew from previous work that the neuropeptides oxytocin and vasopressin played a pivotal part in creating attachment behaviors found in other mammals, notably monogamous prairie voles. Most important, she knew from a 1999 study published in *Nature* that normally nonmonogamous mice could be made to behave more monogamously by injecting them with both vasopressin and a gene that helped bind vasopressin to receptors.[63]

So, with all of this in mind, could a love potion actually be crafted? The answer seems like a straightforward yes, right? To generate sex drive you'd load people up on a mix of androgens and estrogens. To create affection you'd administer doses of dopamine and norepinephrine while simultaneously administering a drug to reduce serotonin availability. Finally, to increase feelings of attachment, you'd deliver a mix of oxytocin and vasopressin.

"The thing is, we've got a list of traits in our heads, a love map of sorts, that controls who we will fall in love with," explained Dr. Fisher. She does not dismiss the idea that sending dopamine levels through the roof will cause temporary attraction. "It definitely will. The catch is that it won't last. In the morning you're going to look at that person and wonder what the hell you were thinking." Even then, she argues, biochemistry has limits. "I can promise you that you could never dope me up enough to fall in love with George Bush!" she laughed.

It made me wonder if the crucial issue was distance on the love map. For Dr. Fisher, President Bush was metaphorically a thousand miles away from the sort of partner that her brain was telling her she should have. But what about someone who was only a few hundred miles away, close friends but not romantically involved? "If you've got one or two of the systems active already, I suspect you could chemically meddle with things and close the gap," explained Dr. Fisher.

This is dangerous business, though. While giving androgens to somebody with whom you have attachment could lead to the person's leaping into bed with you, it could also cause serious problems. While this has not been directly studied in people,* an experiment with birds run by zoologist John Wingfield in 1994 found that when mated male sparrows were injected with testosterone, every single one of them abandoned its partner.[64]

Similarly, another study led by Alan Booth at Pennsylvania State University in 1993 found that men with high baseline levels of testosterone married less often, were more abusive during marriage, and divorced far more often than men with standard levels of baseline testosterone.[65] This suggests that sex drive can, and often does, have a negative impact on feelings of attachment.

As for the reverse, creating a potion that can cancel infatuation or, to put it more magically, "break enchantment," Dr. Fisher says, "We're already there, and it's a good thing too because so many suicides and crimes are committed by people who are absolutely lovestruck." Dopamine, the main compound behind infatuation, and serotonin work in opposition to one another. As one goes up, the other goes down. Thus, if you administer a serotonin booster to someone overwhelmed by infatuation, it can help pull him or her out of it. "In essence, what you are doing is blunting the emotions, and if we were to do that for a person who has fallen in love with their boss who has a spouse and three kids, we'd be doing them a real service. Mind you, the memories would still be there, and those would hurt, but the serotonin would be a real help," Dr. Fisher said.

With so many drugs now known that do many of the things that

---

* There are laws against this sort of thing.

magical enchantments historically have done, it might seem that these forms of magic have lost their place in our world and been replaced by science, but I'd argue that this isn't so.

In the film *Batman Begins*, a major source of Bruce Wayne's power is shown to come from his ability to control and harness his fears.[66] Yes, he is physically strong and well schooled in martial arts, but the power that he learns to hold over his mind is what allows him to become a hero. This mental control is tightly tied to meditation and hallucinogenic drug exposure, but is this science? I doubt it. Nobody has done work showing that combining hallucinogens with meditation and thoughts of one's worst nightmares will make you stronger. If anything, the evidence that we have suggests psychotic episodes are more likely. Nevertheless, as *Batman Begins* pushes pharmacology and psychology to their outer limits, it transforms these sciences into magic that we, the audience, can readily believe in.

Similarly, in *The Matrix* we were presented with the idea that if people could unlearn what they had learned in life, they would be capable of accomplishing incredible feats.[67] Leaving the virtual reality element aside, martial arts experts who learn to understand exactly how their bodies interact with the physical world around them *are* capable of amazing feats.

Is attaining a peaceful state that allows improved performance actually something supernatural? The countless millions who pray or meditate before competitions suggest that this is the case. It is from these beliefs, I think, that we look upon the red pill in *The Matrix* with such fascination. We have already created sleep potions. Love potions are not far off, and drugs that mimic the apple in Eden are already being tested in labs. It all makes me wonder if we will soon be able to take a red pill and see how far the rabbit hole really goes. Or is a pill even necessary? With that question in mind, let us depart the world of enchantments and begin exploration of what it takes to become superhuman.

# SUPERHUMANS

## Savants, Buddhist Monks, Sword Swallowers,
## Firewalkers, and Stage Magicians

*Free your mind.*

—MORPHEUS, *THE MATRIX*

In the film *Rain Man*, Dustin Hoffman plays an autistic savant named Raymond, who demonstrates an exceptional talent for counting.[*][1] But counting is just the tip of the iceberg. Over the years savants have proven themselves capable of instantly spotting prime numbers, knowing which day of the week a date years in the future will fall on, drawing nearly photographic pictures without having had any artistic training, learning foreign languages without having to work at it, and solving Rubik's Cubes in minutes. One autistic boy solved the cube in a minute and seven seconds.[**] Among the most famous of such savants is Daniel Tammet, who memorized pi to 22,514 decimal places, learned

---

[*] The scene in that film when Raymond immediately counts the number of toothpicks that have fallen on the floor in a café mimics an event that actually took place. In 1985, a couple of autistic twins immediately identified the exact number of matchsticks that had fallen on the floor in front of an autism researcher. They said in unison, "One hundred and eleven."

[**] The world record as of 2013 is 5.55 seconds, but I think we all can admit that the autism feat is still pretty amazing, given that the kid didn't look up any of the cheats for solving the cube online and didn't practice.

Icelandic in a week, and can calculate exceptionally complex mathematics at lightning speed.[2]

In 2009, in *Philosophical Transactions of the Royal Society*, Simon Baron-Cohen, a leading expert on autism at the University of Cambridge, explained, "The excellent attention to detail is directed toward detecting 'if p, then q' rules."[3] Said another way, the mind of a savant is outstanding at identifying and exploiting laws found in our world. If we multiply 3 by 4, then we get 12. If we throw a ball in precisely this manner, then it will curve to the left. If it is Tuesday, then we eat Thai food.* Laws, which are always true, are found in numbers and calendars. They are found in optics, which is why savants often demonstrate an uncanny ability to draw. Even languages are mostly systems of coherent rules and logical patterns. Linguistic savants, often called "living Rosetta stones," pick up on these patterns.[4]

It is one thing to say that savants have minds that readily latch on to laws and apply them without breaking a sweat, but another to explain what their brains are actually doing. One theory is that, because their brains function in an atypical manner, they have access to information in a raw form that you and I do not. The argument goes something like this: A normal human mind collects information and immediately starts packaging and labeling it before the information is consciously processed. For this reason we see people, bunnies, taxicabs, etc., as whole units rather than as component parts. In contrast, the minds of savants "see" information before it is processed into a whole. This allows them to spot things that an ordinary human mind would miss. For example, if you or I were shown a photograph of a smiling woman with six fingers on her right hand, we'd see a smiling woman and be unlikely to detect the six fingers unless we studied the photo closely. People with autism would detect the

---

* The writers behind *The Big Bang Theory* have clearly done their autism homework in developing Sheldon's character.

six fingers immediately because they access the visual information before it is packaged by their brains.[5] This ability to readily spot details rather than simply taking in whole images also renders autistics immune to stage magic.[6] It doesn't matter how talented the magician. Copperfield, Blaine, Penn and Teller . . . they would all have a tough time tricking a savant, because their brains do not make the assumptions or take the shortcuts that typical brains do. It makes savants highly resistant to being duped by sleight of hand and excellent at seeing through illusions.[7] So is it possible for ordinary people to bypass the paths typically used by their brains and get at unprocessed information as savants do? Yes.

Autistic savants tend to have left-brain dysfunction coupled with right-brain compensation, and this has led numerous research groups to wonder if sabotaging a portion of the left brain might grant savant-like abilities. Numerous experimenters, but most especially neurobiologist Allan Snyder at the University of Sydney in Australia, have used magnetic pulses to temporarily disable the left anterior temporal lobe of the brain in ordinary people before giving them specific tasks.[8] In one case, participants were given a minute to draw a horse, dog, or face. In others, they were given challenging proofreading or number-estimation tasks after being exposed to the magnetic pulse. In all experiments, a portion of the participants showed dramatic improvements.[9] After one drawing experiment, one man could not believe that the highly accurate drawings were his own. Yet the effects were not universal; savantlike skills were not induced in everyone. Nobody knows why. That's obviously worth looking into, but even more intriguing is what happens to these superhuman abilities when autism is ultimately cured.

## FOLLOWING ALGERNON

In case you think I am just making some sort of wild speculation here about Daniel Keyes's *Flowers for Algernon* coming true, think again. In Keyes's story, a mentally disabled janitor named Charlie is healed of his handicap with an experimental medical procedure shortly after it does the same for a mouse named Algernon.[10] Charlie's tale was the stuff of fiction when it was written in the 1950s, but—based upon work being done in cellular biology—it does not look as if it will remain fiction for much longer.

Robert Naviaux, a cellular biologist at UC San Diego, was fascinated by autistic savants and curious about what precisely was happening in their brains that granted them their incredible abilities. "Neurons that are developing contacts in the brain 'look' for little lights in the darkness, the metabolic activity of other neurons that have been activated by use," explained Dr. Naviaux. Once they find these lights, they send out projections called axons to handshake with them and form connections. "This is why play and communication, and giggling and the recognition of mother's face, are so important in the first year of life. The physical and mental use of neural circuits makes cells 'light up' with metabolic activity so neurons can find one another and make new connections, sometimes at long distances in the brain." Yet, this connection making can go wrong.

"When cells suspect danger, they resist sending out the longer connections. Their axons are shorter and they have fewer branches. However, even though they send out fewer long axons, neurons still engage in local activity as the drive to make connections during early development is very strong," said Dr. Naviaux. The result is a brain filled with isolated islands of superconnectivity that can perform specific tasks at remarkable speeds.

Dr. Naviaux knew that the cellular danger response took effect when cells were "told" by organic compounds in the blood, called purines, to stop performing their ordinary services and activate defensive systems to stave off attack from viruses or toxic chemicals. Moreover, he suspected that when the danger response was activated in young children, the purines sometimes just kept circulating nonstop and left neurons in permanent defensive lockdown. If this was so, then he theorized that targeting the purine activity could potentially end the lockdown, allow the neurons to make long connections again, and treat the autism.

Dr. Naviaux set up an experiment with mice. He knew from past work that if female mice were infected with a virus midpregnancy, their offspring would be born with brains in defensive lockdown and exhibit autistic behaviors such as fear of novelty and difficulty interacting with other mice.[11] A drug called suramin was already on the market for the treatment of the sleeping sickness spread by the tsetse fly in Africa and was known to bind to purine receptors on cells found throughout the body.[12] Dr. Naviaux used it on both autistic and nonautistic mice when they were six months old.* The mice were then tested on how well they interacted with mice they had never before met, and on whether they preferred sameness when put through mazes, as many autistic humans do.

Suramin levels in the blood declined naturally, dropping by half for every week that passed after the initial dose.[13] When five weeks had passed, the mice were put through the tests one final time. Throughout the study, a few mice were sacrificed from both the control and the experimental groups so the biochemistry of their neurons could be analyzed.

Dr. Naviaux and his colleagues reported in *Translation Psychiatry* in 2014 that while some brain cells, called Purkinje cells, that would normally be found in healthy brains did not suddenly reemerge in these autistic mice, suramin ended the cellular defensive lockdown.[14] Moreover, the team discovered major behavioral changes in these animals.

Mice that had initially shown a fear of stranger mice showed no sign of this behavior after they were treated with suramin.[15] They behaved like ordinary mice for as long as the suramin was in their systems. After five weeks, when the suramin was effectively gone, their autistic behavior returned.[16]

Whether blocking purines will have the same effect in people needs to be tested. Moreover, suramin has a long-term toxic effect, so a safer

---

* Which in human years is about thirty years of age.

alternative needs to be found. Nevertheless, that an autistic human could potentially be granted a normal life with a single drug is amazing. Yet, as I worked through Dr. Naviaux's papers one thought lingered: Would an autistic savant want an ordinary life?

If you could learn entire languages in a week, draw perfect pictures without thinking about it, and make complex calculations in your head in seconds, would you be willing to trade that for the ability to comfortably walk into a coffee shop and shake a stranger's hand? My thoughts went racing to modern comic mythology such as *X-Men: The Last Stand*, in which mutants are presented with a drug that could make them normal by robbing them of their powers.[17] Would purine treatment in autistic savants be the same sort of thing? I had to know, so I asked Dr. Naviaux, "Will savants lose their mutant powers if we cure them?"

"Treating an autistic savant with a purine inhibitor should not change his or her extraordinary abilities at all," he said. The hyperconnected islands of neurons that were formed when an autism patient was young would still be there. The powers that the islands grant would still be available, but they would no longer be the only neural connections available. "In younger patients in particular, new connections, longer ones, stand a good chance of forming once the danger response is shut down by a purine blocker," says Dr. Naviaux. The result? A superhuman mind without the burden of autism. Wow.

## MIND OVER MATTER

The German free diver Tom Sietas was born with lungs 20 percent larger than average. He can hold his breath far longer than most people.[18] On a good day, I can sit at the bottom of a swimming pool for three minutes before having to come up for a breath. While I am in relatively good shape and trained as a diver, this is nothing compared to the twenty-two minutes and twenty-two seconds that Sietas pulled off in 2012.[19] How'd he do it? Practice.

Regular breathing exercises that help improve overall lung capacity are important. So too are a healthy diet and good physical fitness so the body can run as efficiently as possible when starved of oxygen. However, for Sietas to hold his breath for such a long time, he also needed to learn how to manage the resources in his body wisely.

Every pump of the heart consumes oxygen, so breath holders must consciously lower their heart rates. Muscles need to be relaxed, and the brain must be calmed, as even thinking drives neural activity that ends up consuming vital oxygen.

As oxygen runs out, the lungs deflate, triggering pain and panic. It also drives the brain to scream for action to be taken. Coping with the pain and ignoring the mental emergency messages demands mastery of meditative practices that help the conscious mind take control of the body as it freaks out.[20] This is challenging. I certainly can't do it, but Sietas and his fellow breath holders have learned to do so. They aren't alone.

Meditation is common before martial arts practices and competitions. It helps to ease the mind and prepare it for rigorous physical activity. One study published in the *British Journal of Sports Medicine* even reported that meditation helped shooters reduce stress and take a more steady aim at riflery targets.[21] However, meditation gets put to the most phenomenal of uses in the hermitages of mountain-dwelling Buddhist monks.

There he sat in the snow. Naked and cross-legged in the midst of winter. The thermometer read 24.8°F (−4°C), but the monk just sat there without showing any signs of suffering. It was, admittedly, just a YouTube video passed along to me by a friend who knew my interest in all things supernatural.[22] Nevertheless, once I saw it, I couldn't stop thinking about it. How could someone sit in the cold like that and not start shivering? How could the person not fall victim to hypothermia? How did his fingers and toes avoid succumbing to frostbite? It was marvelous and perplexing. I had to learn more.

I didn't expect to find anything academic, let alone scientific, on this strange practice. However, as I ran my searches through the journals, I was astonished to discover that I was quite wrong. In the journal *Nature*, of all places, was a paper examining this very thing.[23]

Herbert Benson at Harvard Medical School described activities even more extreme than what I had recently seen on the Internet. He had watched as several monks, wrapped in sheets that had been drenched in water from a nearly frozen well, sat on an icy stone floor one winter's night. It sounds insane, but these men were competing. Each was using his body heat to dry as many sopping sheets before dawn as he could.

These Buddhist monks say the meditation "g Tum-mo" allows them to supernaturally heat their bodies by channeling a mystical wind called prāṇa, which is normally scattered throughout the body. This ignites an internal fire that allows them to warm external portions of their bodies.[24]

Dr. Benson and his colleagues got the permission of the Dalai Lama to closely study three monks living in the foothills of the Himalayas who were experts at g Tum-mo. These monks had lived for at least ten years in near-total isolation in unheated, uninsulated stone huts with earth floors and tin roofs. They had never lost any fingers or toes to frostbite. The researchers attached small temperature-sensing devices to the fingers, toes, forearms, legs, navels, and nipples of the monks. To get a sense of what was happening to their internal body temperatures, the team also arranged for thermometer probes to be stuck ten centimeters beyond the monks' anal sphincters.*[25]

Once all of the sensors were in place, the monks were asked to engage in their usual meditations. All of them demonstrated an ability to raise the temperatures of their fingers and toes while sitting cross-legged and motionless in their Buddhist lotus positions. In some cases the temperature increases were relatively modest. For example, the

---

* No wonder the researchers needed the permission of the Dalai Lama.

monks were usually only able to raise the temperatures of their navels by 3.42°F (1.9°C) and their nipples by 2.7°F (1.5°C). In some parts of their bodies, such as their anal regions, their temperatures changed almost not at all. However, they showed an unbelievable ability to heat their fingers and toes.[26]

All were capable of heating their digits by more than 5.67°F (3.15°C). One of the monks raised his finger temperatures by 12.96°F (7.2°C), and another was able to raise his toe temperatures by a whopping 15°F (8.3°C).[27] That's more than enough heat to protect the skin from frostbite.

The meditation was undeniably releasing heat in the body. Nonetheless, it left Dr. Benson with a huge question about what, biologically, was going on. To date, nobody is entirely sure. Some speculate that the monks have learned to increase their metabolism and produce more body heat, but Dr. Benson suspects it is more likely that the monks have learned to take mental control of their circulatory systems.[28]

Under normal circumstances, the blood vessels along the surface of our bodies constrict when we get cold and dilate when we heat up. This is to help keep our body temperatures stable. By constricting under cold conditions, our surface blood vessels help us to conserve heat in the cores of our bodies. This keeps our brains, hearts, and other vital organs alive at the expense of more disposable bits, such as our fingers and toes. Conversely, the blood vessels near the surface of our skin expand as we get too warm to help us shed heat and prevent our brains from cooking.

Most of us don't have conscious control of our circulatory system. But we know through other studies on meditation that it can be controlled with practice. A 1978 study from the *American Journal of Physiology* found that blood flow could be increased by as much as 44 percent during ordinary meditative practices.[29] Thus blood flow between the warm core and the cooler digits may be increased during g Tum-mo. If this is happening, it is metabolically expensive for the monks, since heat sent to the fingers is ripped away by cold air quite quickly. This heat loss means more heat must be produced by the burning of fuel within the body to prevent hypothermia. I've looked for research on calorie intake by monks who practice this unusual meditation and come away empty-handed, but I'll bet that they are eating a whole lot more than usual when heating their digits in this way.

Buddhist monks are not the only ones who must train long and hard to perform seemingly superhuman acts. People who swallow swords, a practice that has been around since 2000 BC, must also intensively train their minds to overcome the natural responses of their bodies.

To many, sword swallowing seems deceptively simple at first glance. It is just about getting control of that pesky gag reflex and then shoving a sword down your gullet, right? Not so. The process is complex. While the gag reflex is the first involuntary response that a sword swallower must control, it is only one of many obstacles to overcome.

A 2006 study published in the *British Medical Journal* surveyed members of the Sword Swallowers Association International to find out how they swallow swords without injuring themselves.[30] This group* requires people to swallow a solid-steel sword that is at least sixteen inches long, provide their blood type and a doctor's phone number for emergencies, and pay twenty-five bucks to gain membership. They are the real deal, and the best place to find out how sword swallowing actually works.

Information gathered from forty-six members revealed that they usually learned to tame their gag reflex by sticking mundane things such as fingers, spoons, knitting needles, and plastic tubes down their throats regularly for several months before moving on to more challenging objects such as bent coat hangers. Once they mastered this, they had to practice

---

* One hundred nine members and counting.

aligning the point of the sword, which most swallowers lubricate with saliva or butter before sticking the sword in, with their upper esophageal sphincter while their neck was hyperextended. They then had to learn to relax both their pharynx and their esophagus. Since these two bits of the body are not under voluntary control, this is often challenging. Even so, all of the sword swallowers in the group managed it and described the action as something akin to relaxing muscles in the neck. After getting the sword that far down, they had to nudge their heart a bit to the left and sink the sword into the stomach. Having a sword in the stomach causes the retch reflex, which, like the gag reflex, exists to help get foreign material out of the body. This too required the swallowers to remain calm and learn to control the stomach's involuntary reaction.

So that's the science of sword swallowing all summed up rather tidily from the 2006 paper. However, the researchers behind the *British Medical Journal* article were not the first to take an academic interest in this arcane art. That honor goes to the German doctor Adolf Kussmaul, who, aside from pushing medical understanding of diabetes miles forward and discovering dyslexia, helped to develop the modern endoscope in 1868.[31]

To anyone who has ever had an ulcer or suffered from severe heartburn, this device is probably uncomfortably familiar. I know it firsthand because, in 2006, I had an endoscope shoved down my throat after I suffered a period of bad acid reflux. In the USA, I hear that gastroenterologists traditionally give patients sedatives before this procedure. Not so in the United Kingdom. My doctor sprayed the back of my throat with a numbing agent that tasted a bit like banana, then jammed an enormous tube with a camera inside it down my throat. The procedure is well understood today, but back in Dr. Kussmaul's time, navigating a tube through the various sphincters in the throat was uncharted territory. To figure things out, he engaged the services of someone who regularly stuck stuff down his throat: a professional sword swallower.[32] The swallower, whose name has vanished into history, spent a lot of time explaining what he had to do to swallow swords, and this led Dr. Kussmaul to be the first to guide a scope all the way down into a patient's stomach. So, if you think about it, modern endoscopy actually owes a lot of its present knowledge to sword swallowers.

Sword swallowers must practice to do what they do just as breath holders and Buddhist monks using g Tum-mo must practice. However, there is a difference between these groups. Monks and breath holders don't commonly perform their feats in front of audiences. Even when they do, they are not performance artists in the classic sense. In contrast, sword swallowers are more like traditional performers, and for those who make a living swallowing swords, it is in their best interest to put on a good show so they generate strong audiences.

What does putting on a good show entail? It can mean performing advanced sword-swallowing techniques such as "the drop," where a sword is literally dropped down the throat rather than slowly slid into place.[33] The drop technique is all about skill and practice. A good performance might also require making maneuvers that are easy for the sword swallower look challenging to the audience so they are more readily impressed. It is absolutely in the sword swallower's interest to play up the risks and heighten suspense. In short, an element of deception may increase the drama.

## MISCHIEF AND TRICKERY

Fire walking goes back thousands of years. Priestesses of a now-forgotten Asiatic goddess once walked barefoot across burning charcoal in the Greek town of Hierapolis, which is in what is now southern Turkey.[34] At the Roman sanctuary of the fertility goddess Feronia, men demonstrated the powers of their faith by fire walking once a year in front of huge audiences.[35] We even find evidence of fire walking in the Old Testament, with a passage in Ezekiel 28 where King Tyre walks on red-hot coals.[36]

Today, fire walking is practiced worldwide, from the tribes of the Kalahari Desert and the clans living on the Fijian Islands to the Hindus of southeastern Asia and the Eastern Orthodox Christians of Greece and Bulgaria.[37] With the red glow and heat, it is almost impossible to watch someone fire-walk barefoot and not be impressed. Even so, not all is as it seems.

First, consider when fire-walking events happen. We never see them in the morning or afternoon. They are always at night. This is because fire walking, contrary to its name, is not the practice of walking on fire. It is the practice of walking on coals, and coals look dusty gray in daylight because they are covered with ash.[38] However, by night their red glow seeps through the ash, making them appear ominously red.

The second illusion at work is the apparent danger of the coals. Walking on an actively burning fire would leave anyone injured. However, once a fire has been reduced to coals, what remains in abundance is carbon.[39] Carbon is a poor conductor of heat, and ash does not conduct heat well at all. In fact, ash is such a good insulator that it was once used to line the interior of iceboxes to keep heat out of them. This does not mean you can stand on glowing coals coated with ash and not get burned. Stand there long enough and heat will eventually transfer to your skin and cause burns. Fire walkers get around this by moving. They never hang around in one spot long enough for the heat inside the coals to transfer to their flesh.[40] While running might seem sensible, since it reduces contact time with the heat, it is actually a bad move, as it forces the foot through the insulating layer of ash and deeper into the coals.[41] Additionally, running would look cowardly and dispel the majesty of the show.

So does fire walking require some degree of skill? Yes. You need to know when a fire is suitably "ready" for a walk to commence, and you need courage to do the seemingly unthinkable. Beyond that, though, fire walking is a partial illusion. There is a substantial risk, since the insulating layers are not perfect and bad luck can lead to burns. Nevertheless, audiences that believe something magical is being undertaken are being tricked. Indeed, in the Old Testament Ezekiel 28 passage that I mentioned earlier, God is angry at King Tyre for pretending to be a deity and deceiving viewers into believing he has supernatural powers that allow him to walk across fire.

The same sort of trickery is practiced by performers who lie on beds of nails. While lying on a single nail or just a small patch of nails would definitely cause significant injury, lying on thousands of nails distributes a performer's weight across so many points that in no single location is the pressure great enough for the points to pierce the flesh. Some skill is required here too, particularly in lying down and getting up, since performers risk getting hurt during these times when only small portions of their body rest on the nails. Nevertheless, in bed-of-nails performances audiences are being duped into believing the human body ought to be pincushioned when physics dictates otherwise.

These deceptions are nothing compared to those accomplished by master deceiver Harry Houdini. Pegged as the greatest magician of all time, Houdini was an escape artist above all else. During his career, he got police to lock him up using devices that they believed were entirely escape-proof. Manacles, shackles, chains, ropes, straitjackets, metal-lined cells inside prisoner-transport vehicles, with strip-searching—police tried everything, and Houdini always got free.[42]

Over the years he appeared to grow more daring. At some events he was bound and then escaped while hung upside down from a rope. At others he was submerged in a river while handcuffed and locked in a box or he was flooded while chained inside a milk canister. How'd he do it? Good question.

Before he died, Houdini claimed that many handcuffs and locks could be opened with the appropriate application of force.[43] He also explained in notes that he wrote for the wider magic community that he would swallow and then later regurgitate small lockpicks that allowed him to open the devices binding him.[44] When he was tied up with ropes or held fast by a straitjacket, he would dislocate his shoulders to wriggle free.

It all sounds plausible and could be the truth, but there is a problem in that all of these explanations come from the same man who said, "If I got out too quickly, the audience would reason that escape was easy. Every second that ticks by during my struggle builds up the climax. When they are sure I am licked, that the box will have to be smashed open to give me air, then, and only then, do I appear."[45]

In London in 1904, a British workman who had labored for five years to build "pick-proof" handcuffs approached Houdini and dared him to try to escape from the manacles.[46] Houdini wavered at first but soon accepted the challenge in front of a huge audience. He struggled behind a curtain to release himself from the shackles, and after an hour and ten minutes, he succeeded.[47] Understandably, his success was met with roars of approval from the crowd. Was it a deception?

Jim Steinmeyer, a master designer of stage illusions, argues that it was. In his book *Hiding the Elephant*, Steinmeyer points out that the many experts who have examined both the handcuffs and the records of the event have concluded that Houdini staged the whole thing.[48] Houdini had requested that the shackles be made. He had also arranged for these shackles to be passed along to the man who ultimately challenged him. As Steinmeyer puts it, "The performance was a glorious deception."[49] It was stage magic.

## SMOKE AND MIRRORS

A white goose walked onto the stage. David Copperfield grabbed it, stuck it in a box, and then disassembled the box. The goose was gone. As everyone looked on in amazement, he revealed the bird, hiding in a bucket that the magician had given an audience member to hold just minutes earlier. This impressive trick, like all such tricks, got my brain chewing. Had the bucket really been empty? Had the audience member been planted by the magician? Had the goose been transported, or did our undiscerning human eyes assume two geese were the same animal because they looked alike?

It was all a bit depressing. After spending almost two years wading through the arcane practices of our ancestors and the brilliant ways in which modern research teams are making the magic of our myths spring to life, the tricks of Las Vegas felt hollow. Card forces, coin tossings, vanishing geese—could I even classify this stuff as magic? Was there any science behind it? Kind of.

During a late evening at the Rio I met with Sebastian Walton, an eighteen-year-old Briton who won the Young Magician of the Year award for 2014. As we chatted about how he got into magic, I watched him perform a card trick he had just invented in front of a couple of elderly magicians. They were polite and lauded Sebastian's work, but they were also quick to point out how he could use age-old tactics to make his moves even more deceptive. In an emotional moment for me I watched as an aged but agile hand reached out for the cards to demonstrate a more convincing mechanism for palming.

Young magicians are not the only ones studying old material and experimenting with variations upon it. At the opening of their show in Las Vegas, Penn and Teller asked for a volunteer with a smartphone to come up onstage. A gorgeous lady in a gold dress obliged and handed Penn her phone. He made to put it into an empty coffee cup but, behind the lady's back, tossed the phone in full view of the audience to Teller, who dumped it in a red bucket hanging from a rope. Then, to the lady's surprise, Penn stomped on the cup with his foot. When she surveyed the wreckage, no phone, or pieces of a phone, were to be found. Penn advised that she might be able to find her phone if she asked her boyfriend, a guy in the audience, to give it a call. When he did, a ringing emerged from the thirteenth row of the theater. Guests started searching around and, under a seat, they discovered a Styrofoam box. This box was passed to the front, where it was opened to reveal a ringing fish on ice. Teller chopped off the fish's head and revealed the phone, sealed in a plastic bag inside the animal.

Later that evening, backstage in Teller's sitting room, the magician snatched my notepad and started looking over the scribbles I'd made during the show.

"No. That's not how we did that. You think I did what? Really? You ought to try that and see for yourself that it wouldn't work." It was cringe-worthy stuff. I could readily propose theories for the science behind liver readings, ancient healing rituals, and the secret of the Mayan blood rites, but I knew next to nothing about the sort of magic that Teller was performing. He looked me hard in the eye and said, "You need to learn more about magic." He scribbled down book titles in my notepad.

Two days later I was back in London and browsing the tomes he'd referred me to. Some of these things were seriously old. Several, to my surprise, were in the rare-books room of the British Library. One, published in 1894, was yellowed with age, falling apart, and held closed by a bit of silver cord lovingly tied into an elaborate knot. It looked like an ancient spell book, and as I read it, I realized this was not far from the truth.

### The Watch Mortar and the Magic Pistol

*Borrow a watch from someone and suggest to the owner of the borrowed time-keeper that it wants regulating and offer to undertake that duty for him. He probably declines, but you take no notice of his remonstrations and, placing his watch in the mortar, bring down the pestle with a heavy thump upon it. A smash, as of broken glass, is heard and after sufficient pounding, you empty the fragments of the watch in your hand to the horror of the owner. You offer to return the fragments but he naturally objects to receive them and insists that you restore the watch in the same condition as when it was handed to you. After a little discussion, you agree to do so, promising that you can only effect the object through the agency of fire. Fetching a loaf of bread,*

*you place it on the table in view of the company. Then wrapping the frag-*
*ments of the watch in paper you place them in a pistol and aiming at the loaf*
*request the owner of the watch to give the signal to fire. The word is given*
*one, two, three . . . bang! Stepping up to the loaf you bring it forward to*
*the spectators and tearing it asunder exhibit in its very centre the borrowed*
*watch completely restored and bright as when it first left the maker's hands.*[50]

Sound familiar? Swap a smartphone for the watch, the coffee cup for the mortar, Penn's foot for the pestle, and the fish for the loaf of bread, and you've got the same trick. Even so, special attention must be given to the addition of the fish.

In our myths, fish have a long history of retrieving valuable objects from the sea. The tale of Abu Sir the Barber, in *The Arabian Nights*, describes how a fish brought the impoverished barber the king's signet ring: "Then he cut its throat with a knife he had with him, but the knife stuck in its gills, and there he saw the King's signet ring, for the fish had swallowed it and destiny had driven it to that island, where it had fallen into the net."[51]

The old tale of Abu Sir the Barber was translated into English by Sir Richard Burton in 1850 and is thought to stem from oral stories that arose in medieval Persia. However, it may stem from something even older, because the Greek historian Herodotus reported a similar story in 440 BC about a king named Polycrates whose signet ring ended up being returned to him inside a fish: "The servants, cutting up the fish, found in its belly Polycrates' seal-ring. As soon as they saw and seized it, they brought it with joy to Polycrates, and giving the ring to him told him how it had been found."[52]

Is this one myth that got repeated and shaped over time in the Abu Sir story or did it arise independently in numerous locations? A search through the *Motif-Index of Folk-Literature* suggests the latter, as dozens of stories appear from around the world about fish recovering valuables associated with identity—signet rings and the like—from the sea.[53]

Perhaps the story naturally formed among people who fished every day and has stuck with us over time? Teller agrees: "If you spend every day pulling fish from the ocean and cutting them open, I can't imagine *not* fantasizing about something valuable inside!"

Penn and Teller were not the first magicians to incorporate fish into their act. That honor goes to the father of modern magic, Robert-Houdin, who lived from 1805 to 1871. How does the ring, watch, or phone end up in the fish or loaf of bread? Avert your eyes if you don't want to know, because here are the details:

> *The seeming mystery is easily explained. The mortar has a moveable bottom which allows the watch at the performer's pleasure to fall through into his hand. There is a hollow space in the thick end of the pestle, closed by a round piece of wood lightly screwed in, which, fitting tightly in the bottom part of the mortar, is easily unscrewed by the performer or rather unscrews itself as he apparently grinds away at the ill fated chronometer. In the cavity are placed beforehand the fragments of a watch, which, thus released, fall into the mortar and are poured out by the performer into his hand in order to show that there has been "no deception." When the performer goes to fetch the loaf, he has already obtained possession of the watch, which, after giving it a rub upon his coat sleeve or a bit of leather to increase its brightness he pushes into a slit already made in the side of the loaf. When the loaf is torn asunder (which the performer takes care to do from the side opposite to that in which the opening has been made) the watch is naturally found imbedded therein.*[54]

Explanations like this make stage magic seem simplistic, but this simplicity is itself an illusion. "Magicians guard an empty safe," writes Steinmeyer.[55] What he means is that the science behind stage magic is relatively simple. The real skill of the magician is wielding these simple concepts and tools to make audiences believe they are seeing the impossible. Robert-Houdin described his role as "an actor playing the part of a magician."[56] This suggests that magicians are actors, and that, since acting is an art, so too is stage magic. I mostly agree with this sentiment but feel this analysis ignores the process of creating new tricks.

During another of our chats, Teller described how he devised one trick while having lunch in a café by crumpling up paper napkins and methodically studying how pouring them out of an empty cup in various ways drew his gaze differently. He wasn't working with a large population. He didn't have a control per se. He wasn't even taking notes, yet this was experimentation nevertheless. The biggest secret behind magic is that all the monotonous effort that goes into testing and creating it is ugly, he explained.

What I found most striking about Teller's statement was how true it was of the research world as well. While studying in the paleontology labs at UC Berkeley, I participated in several research projects and contributed to a few academic papers.[57] One project used a bunch of mammal fossils found in cave sediments to figure out what the climate had been like when the mammals had all been alive. I must have spent over 250 hours staring at rodent teeth under the microscope to write my measly four-paragraph paleoclimate section for that project's paper, and that is nothing compared to what most in the scientific community must endure before getting their works published. Talk about ugly.

Penn hammered the similarities home even harder: "We tried this walking-on-water trick with a fire hose. God, what a fucking mess. After trying over and over again, we got it to work, but the back spray was a nightmare. It went everywhere. We had to put screens up to control it, and they just ruined the whole thing; you couldn't see anymore. After spending I don't know how many grand, we had to ditch the whole idea." As I listened to his story, my mind wandered back to a scientist whom I had chatted with just a few weeks earlier. She was justly proud of a recent publication in the journal *Nature*, and while I was interviewing her, she sighed, saying, "After years and years of unpublishable findings it is such a relief to finally be at this point." Just like the scientists, Penn and Teller had their version of unpublishable findings.

Could the devising of magic tricks really qualify as science? Science, at least according to the dictionary on my desk, is "a systematic enterprise that builds and organizes knowledge in the form of testable explanations and predictions about the universe."[58] We don't think of magicians as publishing their work, but as the musty tomes at the British Library make clear, they do, albeit rather quietly. Thus, magic is an enterprise that builds and organizes knowledge. Magicians are even forming testable explanations and predictions about the universe.

Consider when Teller was pouring crumpled napkins in the café. His idea was that pouring in a specific manner would draw the human eye strongly enough so that other actions happening at the same time would not be noticed. If you think about it, he had built a hypothesis that day in the café about how the human mind would behave, and when he presented it to audiences at the Rio, he was testing out the theory to see if audiences would fall for it. The same can be said for Copperfield's goose and the watch in the 1894 tome. Some magician, long ago, theorized that the human mind would readily latch onto continuity, e.g., the goose in the bucket *is* the one that was in the box, and the watch bits in the mortar *were* once the surrendered watch, even if the continuity depends upon magic that we know can't be true as an explanation. When that theory was tested on viewers and proved effective, the finding was recorded.

I am not the first to decide that magicians have been rigorously studying human behavior for centuries and quietly passing their findings along. Numerous neurologists and cognitive psychologists over the years have studied magicians with the intent of tapping into their age-old techniques and using this information to better understand how the brain operates. These studies have often gained a lot of media attention, because, as Teller put it during one of our conversations, "Magicians are sexier than lab rats"—but some of the work has revealed genuinely interesting results. Top among these was one study conducted with a professional rogue.

## MAGIC UNDER THE MICROSCOPE

Apollo Robbins, best known as the Gentleman Thief, is a magician who specializes in sleight of hand and stage pickpocketing. During his years of training and performance, he found that he could draw a target's attention in different ways by adjusting how he moved his hands. More specifically, if he captured a victim's attention with one hand and moved that hand in an arc, it would draw a victim's eyes differently than if he moved his hand in a straight line.

Intrigued by this phenomenon, a couple of neuroscientists, Susana Martinez-Conde and Stephen Macknik at the Barrow Neurological Institute, decided to take a closer look, being card-carrying magicians themselves.* The two researchers knew the trick of showing a penny between the fingers and thumb and then moving it close to the other hand so that it looked as if the penny were transferred.[59] Then the magician opens the second hand so that viewers believe the penny has vanished, when it was actually not there in the first place. Known as the French Drop, the trick is learned by every novice magician. The two variations—one in which the coin is taken away along a linear path, and one in which it is taken along a curved path—affect audiences differently.

Neuroscientists watched Robbins perform the French Drop in the two different manners while audience members had their eye movements monitored. Robbins speculated, from his years of experience, that when audience members' eyes followed the linear movement of the hand

---

* Magicians do carry cards. To get them, they need to pass performance tests at professional institutes such as the Magic Castle in Hollywood and the Magic Circle in London, where their peers determine whether they are competent enough to call themselves magicians.

in which they thought he was carrying the coin, they would quickly dart back to the hand where the coin had once been after the straight motion came to an end. In contrast, he guessed that when their eyes followed the curved motion of his hand, their gaze would remain upon that hand for a lot longer. This was exactly what the researchers and Apollo reported in the journal *Frontiers in Human Neuroscience* in 2011.[60]

The data revealed that people process a hand moving in a curved fashion using slow eye movements that neuroscientists call smooth pursuit, while we process a hand moving in a straight line using quick scanning movements of the eyes known as saccades. Each participant watched the French Drop fourteen times, and their eye movements did not change between their first time and their fourteenth.

For stage magicians, this finding has all sorts of implications. The first is, if a magician wants eyes to remain on a fixed point at the end of a hand motion, that hand motion should be curved rather than linear. The second is, at least with regard to the French Drop, the classic maxim of "Never perform the same trick twice" is not true. Our eyes are remarkably obstinate regardless of what our brains actually know from seeing the magic trick performed over and over. All pretty cool, but I think the findings are most exciting when viewed through an evolutionary lens.

For a moment consider the discovery as a paleontologist might. What reasons might there be for our eyes to so stubbornly follow curved movements with smooth pursuit and linear movements with the darting saccades? Well, if warfare functioned as a force for natural selection—and we are pretty sure that it did, since humans have been fighting one another for millennia—then these different ways of watching movement would be extremely valuable on the battlefield.

Here's how it would work. You are an ancient human living sixteen thousand years ago. Your clan is clashing with another clan over access to resources, such as a good coastal spot to collect shellfish or some berry bushes. This conflict escalates into violence, and you are sent, along with many others, by your tribal elders to attack the other tribe. Upon attacking, one of the members of the opposing tribe throws a rock at your head and misses. Once the rock has shot past you, keeping an eye on where it fell is of little relevance to your survival. In contrast, having your eyes dart back to where it came from so you don't get killed by rock number two is essential. So, if some in the human population had saccadic eye responses to rocks being thrown at their heads, and some had smooth-pursuit eye responses, people with saccadic responses would most commonly survive battle and ultimately dominate the human population.

In another paper, the same team of neuroscientists tested how a magician's gaze could control his audience. They followed Mac King, the Las Vegas master of comedic magic.[61]

Mac speculated that while performing a fake coin toss—where he made it look as if he were throwing a coin from one hand to the other—he could increase his viewers' belief in the trick by guiding them with his gaze. He sensed that if he gazed at the hand that was catching nothing, he could enhance the illusion that the coin was actually going there.

To test this out, the research group recorded Mac's toss with a camera. In one instance he performed the trick as it was meant to be performed, with the coin remaining in the throwing hand. In another instance, he actually threw the coin. Sometimes he followed the fake coin toss with his gaze, and sometimes he did not. These videos were then shown numerous times to study participants with Mac's face covered up by a black square in half of the viewings. The participants were then asked whether Mac had actually tossed the coin.

The team found that the coin toss trick was remarkably robust. Viewers needed to watch it around ten times before they started to get better at distinguishing between a fake toss and a real one. However, unlike the experiment with Apollo Robbins, where the magician's sense of how his hand movements drew the eyes of the audience in different ways proved correct, the work with Mac found the opposite. Mac's gaze seemed to have little if any effect. Viewers fell for the fake coin toss regardless of whether his gaze was luring them in the right direction or not.

It is intriguing to speculate how the findings in the Mac King paper might be interpreted from an evolutionary perspective. Dr. Macknik argues, "When interacting with other humans, it is the position of their hands that is of utmost importance, even more than their gaze position." Or, to put it another way, natural selection has driven our eyes to follow the hand with the knife because the knife will kill you, not the eyes.

I would never have thought that a stage magician tossing a coin at Harrah's could shed light on my own field of evolutionary biology, but there it was in Dr. Martinez-Conde and Dr. Macknik's research. The science of magic indeed.

# CONCLUSION

*For this is what your folk would call magic, I believe; though I do not understand*
*clearly what they mean.*

—GALADRIEL, *THE FELLOWSHIP OF THE RING*

As she poured out the water for Frodo to gaze into and see the future, Galadriel expressed bewilderment over why the hobbits called her practice "magic" in the same way that they called the sorcery wielded by the dark lord Sauron "magic." To her, the two were utterly different, yet to the hobbits they were one and the same.[1] It baffled her, and when I read *The Fellowship of the Ring* for the first time more than twenty years ago, Galadriel's words baffled me too. Not so now.

After exploring the magic that our ancestors believed in, the magic that we have written into our stories through the ages, and the magic that we see onstage today, I have come to understand what Tolkien meant.

The ancient magic of our ancestors often depended upon their seeing things or doing things that defied their understanding of the world but demanded an explanation nonetheless. Under these circumstances, magic in the form of sorcery, gods, and other supernatural elements was used to fill the gap and make sense of otherwise confusing things. Magic was used to *explain* the seemingly impossible. To an extent we still see magic used in this way when people claim to have been healed by prayer or to have visited heaven for a brief time while having a near-death experience, but this type of magic is less common than it once was.

More widespread today is magic in the form of stage illusions in Las Vegas, wands at Hogwarts throwing colorful bolts on-screen, and haunted rooms stretching in theme parks. This magic is used to *present* the seemingly impossible as if it were real. It is quite a different thing from what our ancestors called magic. However, this "modern" magic is not so different from the magic that we find in historical literature, where wishes and dreams were depicted through fiction. The key difference is that technology developed for the stage, screen, and theme park now allows what could once only be written about to take shape right before our eyes.

Like naive hobbits talking about the magic of the elves and the magic of Sauron as one and the same, it is easy for us to forget that there are these two fundamentally different magical species. Yet, there is much more to it than that.

Our ancestors had a magical worldview. Just as it was for Galadriel, everything around them pulsed with the supernatural. For the Viking berserkers, no biochemistry was behind their rage; Odin was casting a spell. For pilgrims visiting the oracle at Delphi, no ethylene vapors were creeping up the fault below the temple; the spirit of Apollo was coming to force the priestess to speak with his voice. For those given love potions during the Middle Ages, no botany was involved; it was sorcery. For better or for worse, our ancestors could never leave the magic behind. This is not the case today.

Most of us now only encounter magic when we want to. We choose to see it at the cinema when we want to be blown away by the superpowers of the X-Men. We choose to imagine it when we read about the spells cast by Harry Potter. We choose to believe in it when we walk into a theater where we know a magician is going to trick us. In all of these circumstances, we decide to leave our disbelief at the door and enjoy the wonder that we experience when we think about or observe the seemingly impossible.

This ability to suspend disbelief is important regarding the question of whether knowing how something magical works shapes our experience of it. True believers never doubt that the gods or sorcery are behind the inexplicable things they are seeing. For people suspending their disbelief, this is not the case. No matter how much they set their disbelief aside, no matter how much they truly want to do this, at their core they are nonbelievers. They will always have a lingering awareness that a nonsupernatural explanation exists for what they are witnessing. This might make it sound as if the experience of the nonbeliever is somehow less than that of a believer. I don't think this is so.

With a lot of stage magic it is vital for audiences to have a lingering awareness that what they are seeing is not actually real so they can experience something of a mental burn as what they see clashes with what they know to be true. Playing with the puzzles that magicians present to us and pondering the incredible things we see to try to figure out how they are accomplished is pleasurable.

As with stage magic, we find pleasure in trying to tease out which natural phenomena were recorded in the magical tales of our ancestors, and how the magic in our myths can be made into reality by science. Indeed, our love of this problem solving stands as a testament to the powerful human yearning to explore and explain the world.

With this final point in mind, I'd argue that science and magic are not as much at odds with each other as we tend to think. I might even describe the experience of discovering the science behind our myths as magical.

# ACKNOWLEDGMENTS

This book was very much a team effort. The support of my editors, Brant Rumble and John Glynn, at Simon & Schuster was a real help. So too was the support of my agent, Daniel Greenberg, and his assistant, Tim Wojcik.

Even with all the publishing help that I had, exploring mythology and science side by side is a terrible challenge at the best of times, and I needed a lot of outside assistance. For this reason I owe many experts my thanks.

I am indebted to Christian Amatore, David Ashen-Garry, Brock Bastian, Daniel Blumfield, Jelle de Boer, Edward Boyer, Nicole Burgoyne, Bruce Carnes, Derek Collins, Richard Cowen, George Daley, David Dosa, Michael Rodegang Drescher, Denise Faustman, Eli Finkel, Helen Fisher, Barbara Fredrickson, Alan Gadian, Charles Grob, Sarah Hannis, Harry Hemmond, Eli Herscher, Irene Hu, Ronald Jenner, David Johnson, Jerome Kasparian, Sherry Lentz, Daniel Lieberman, Valter Longo, John Lovick, Stephen Macknik, Chris Mark, Susana Martinez-Conde, Adrienne Mayor, William Milliken, Stephen Mitchell, Robert Naviaux, David Nichols, Jay Olshansky, Donald Pfister, Olivier Rabin, Elizabeth Repasky, Bjorn von Reumont, Bruce Rideout, Guy Ropars, Christopher Rose, Mike Rumsey, Joseph Schneider, Jeffrey Shaman, Barbara Sivertsen, Norman Spack, Daniel Stahler, Elly Tanaka, Teller, Dina Tiniakos, Sebastian Walton, Caroline Watt, Martin Wikelski, Alan Woolf, and Richard Wrangham for providing advice, guidance, and fact-checking.

I owe Ian Cheney; Olga Dobrovidova; Giovana Girardi; Ailie Kerr; Maggie Koerth; Wade Roush; my partner, Thalia; and my parents a great deal of appreciation for carefully reading the manuscript when it was still in rather rough shape and for providing me with such valuable feedback.

I am grateful to the Knight Science Journalism Fellowship at MIT for funding some of my more harebrained plans, and to the Folklore and Mythology undergraduates and faculty at Harvard for helping me to explore some of my crazy ideas more fully. A special thank-you also goes to my good friend Charlie for helping me to make many of my adventures happen. I've got a sunstone and a dinghy with your name on it whenever you're ready to make that voyage to Iceland!

# NOTES

**INTRODUCTION**

1. H. C. Kee, *The Cambridge Annotated Study Bible: New Revised Standard Version* (Cambridge, UK: Cambridge University Press, 1993), 58.
2. *The X-Men* (New York: Canam Publishers Sales Corp., 1963), no. 141.
3. *Amazing Fantasy* (New York: Atlas Magazines, 1962), no. 1.
4. "Ebola's Patient Zero: The Child at the Epidemic's Start," *Frontline*, PBS, December 29, 2014, https://www.youtube.com/watch?v=bDCsvuZhnB8.
5. G. Ropars et al., "A Depolarizer as a Possible Precise Sunstone for Viking Navigation by Polarized Skylight," *Proceedings of the Royal Society A: Mathematical, Physical and Engineering Sciences* 468, no. 2139 (2011): 671–84.
6. A. Plaitakis and R. C. Duvoisin, "Homer's Moly Identified as *Galanthus nivalis* L.: Physiologic Antidote to Stramonium Poisoning," *Clinical Neuropharmacology* 6, no. 1 (1983): 1–5.
7. P. Jackson, *The Lord of the Rings: The Fellowship of the Ring* (New Line Cinema, 2001).
8. A. F. Aveni, *Behind the Crystal Ball: Magic, Science, and the Occult from Antiquity through the New Age*, 1st ed. (New York: Times Books, 1996).
9. B. L. Fredrickson et al., "A Functional Genomic Perspective on Human Well-Being," *Proceedings of the National Academy of Sciences USA* 110, no. 33 (2013): 13684–89; A. Mayor, *The Poison King: The Life and Legend of Mithradates, Rome's Deadliest Enemy* (Princeton, NJ: Princeton University Press, 2010); and K. Ashley, D. Cordell, and D. Mavinic, "A Brief History of Phosphorus: From the Philosopher's Stone to Nutrient Recovery and Reuse," *Chemosphere* 84, no. 6 (2011): 737–46.
10. S. L. Macknik, S. Martinez-Conde, and S. Blakeslee, *Sleights of Mind: What the Neuroscience of Magic Reveals about Our Everyday Deceptions*, 1st ed. (New York: Henry Holt, 2010).

**CHAPTER ONE: HEALING**

1. J. Jouanna and P. van der Eijk, *Greek Medicine from Hippocrates to Galen: Selected Papers*, Studies in Ancient Medicine (Leiden, Netherlands: Brill, 2012), 103–4.
2. Ibid.
3. H. C. Kee, *The Cambridge Annotated Study Bible: New Revised Standard Version* (Cambridge, UK: Cambridge University Press, 1993).
4. S. Cohen et al., "Positive Emotional Style Predicts Resistance to Illness after Experimental Exposure to Rhinovirus or Influenza A Virus," *Psychosomatic Medicine* 68, no. 6 (2006): 809–15.
5. A. Steptoe et al., "Neuroendocrine and Inflammatory Factors Associated with

Positive Affect in Healthy Men and Women: The Whitehall II Study," *American Journal of Epidemiology* 167, no. 1 (2008): 96–102.

6. J. K. Boehm and L. D. Kubzansky, "The Heart's Content: The Association between Positive Psychological Well-Being and Cardiovascular Health," *Psychological Bulletin* 138, no. 4 (2012): 655–91.

7. B. E. Kok and B. L. Fredrickson, "Upward Spirals of the Heart: Autonomic Flexibility, as Indexed by Vagal Tone, Reciprocally and Prospectively Predicts Positive Emotions and Social Connectedness," *Biological Psychology* 85, no. 3 (2010): 432–36.

8. Ibid.

9. Ibid.

10. B. E. Kok et al., "How Positive Emotions Build Physical Health: Perceived Positive Social Connections Account for the Upward Spiral between Positive Emotions and Vagal Tone," *Psychological Science* 24, no. 7 (2013): 1123–32.

11. Ibid.

12. Ibid.

13. Ibid.

14. Ibid.

15. B. L. Fredrickson et al., "A Functional Genomic Perspective on Human Well-Being," *Proceedings of the National Academy of Sciences USA* 110, no. 33 (2013): 13684–89.

16. Ibid.

17. Ibid.

18. Ibid.

19. B. E. Kok and B. L. Fredrickson, "Evidence for the Upward Spiral Stands Steady: A Response to Heathers, Brown, Coyne, and Friedman (2015)," *Psychological Science* 26, no. 7 (July 2015): 1144–46; B. L. Fredrickson, K. M. Grewen, S. B. Algoe, A. M. Firestine, J. M. G. Arevalo, J. Ma, and S. W. Cole, "Psychological Well-Being and the Human Conserved Transcriptional Response to Adversity," *PloS One* 10, no. 3 (March 2015).

20. Kee, *The Cambridge Annotated Study Bible*.

21. G. Pinch, *Handbook of Egyptian Mythology*, Handbooks of World Mythology (Santa Barbara, CA: ABC-CLIO, 2002).

22. I. Tapsoba et al., "Finding Out Egyptian Gods' Secrets Using Analytical Chemistry: Biomedical Properties of Egyptian Black Makeup Revealed by Amperometry at Single Cells," *Analytical Chemistry* 82, no. 2 (2010): 457–60.

23. Ibid.

24. Ibid.

25. A. S. Mercatante, *Who's Who in Egyptian Mythology* (London: Scarecrow Press Inc., 1995).

26. Ibid.

27. Tapsoba et al., "Finding Out Egyptian Gods' Secrets Using Analytical Chemistry."

28. J. P. Bressler and G. W. Goldstein, "Mechanisms of Lead Neurotoxicity," *Biochemical Pharmacology* 41, no. 4 (1991): 479–84.

29. Ibid.

30. Ibid.

31. Ibid.

32. Ibid.

33. Ibid.

34. http://www.groundwateruk.org/UK_thermal_springs.aspx.

35. A. F. Aveni, *Behind the Crystal Ball: Magic, Science, and the Occult from Antiquity through the New Age*, 1st ed. (New York: Times Books, 1996). 31–33

36. R. Jackson, "Spas, Waters, and Hydrotherapy in the Roman World," *Journal of Roman Archaeology* 37 (1999): 107–16.

37. Ibid.

38. K. M. Kokolus et al., "Baseline Tumor Growth and Immune Control in Laboratory Mice Are Significantly Influenced by Subthermoneutral Housing Temperature," *Proceedings of the National Academy of Sciences USA* 110, no. 50 (2013): 20176–81.

39. Ibid.

40. Ibid.

41. Ibid.

42. Ibid.

43. Ibid.

44. Ibid.

45. Ibid.

46. Ibid.

47. Hesiod and Homer, *Hesiod, the Homeric Hymns, and Homerica*, trans. H. G. Evelyn-White, Loeb Classical Library (London: W. Heinemann, 1914; New York: Macmillan, 1914), lines 507–44.

48. Ibid.

49. Ibid.

50. D. G. Tiniakos, A. Kandilis, and S. A. Geller, "Tityus: A Forgotten Myth of Liver Regeneration," *Journal of Hepatology* 53, no. 2 (2010): 357–61.

51. Pausanias et al., *Pausanias' Description of Greece* (London: W. Heinemann; New York: G. P. Putnam's Sons, 1918), Phocis, chap. 11, sec. 1.

52. Homer, *The Odyssey*, trans. A. T. Murray, Loeb Classical Library (London: W. Heinemann; New York: G. P. Putnam's Sons, 1919), bk. 10, lines 208–301, and bk. 11, lines 567–600.

53. G. Higgins and R. Anderson, "Experimental Pathology of the Liver I: Restoration of the Liver of the White Rat Following Partial Surgical Removal," *Archives of Pathology* 12 (1931): 186–202.

54. E. M. Tanaka, "Regeneration: If They Can Do It, Why Can't We?" *Cell* 113, no. 5 (2003): 559–62.

55. M. Kragl et al., "Cells Keep a Memory of Their Tissue Origin during Axolotl Limb Regeneration," *Nature* 460, no. 7251 (2009): 60–65; E. Nacu et al., "Connective Tissue Cells, but Not Muscle Cells, Are Involved in Establishing the Proximo-Distal Outcome of Limb Regeneration in the Axolotl," *Development* 140, no. 3 (2013): 513–18; T. Sandoval-Guzman et al., "Fundamental Differences in Dedifferentiation and Stem Cell Recruitment during Skeletal Muscle Regeneration in Two Salamander Species," *Cell Stem Cell* 14, no. 2 (2014): 174–87; and E. M. Tanaka and P. W. Reddien, "The Cellular Basis for Animal Regeneration," *Developmental Cell* 21, no. 1 (2011): 172–85.

56. N. Shyh-Chang et al., "Lin28 Enhances Tissue Repair by Reprogramming Cellular Metabolism," *Cell* 155, no. 4 (2013): 778–92.

## CHAPTER TWO: TRANSFORMATION

1. Homer, *The Odyssey*, trans. A. T. Murray, Loeb Classical Library (London: W. Heinemann; New York: G. P. Putnam's Sons, 1919), bk. 10, lines 208–301, and bk. 11, lines 567–600.
2. Ibid.
3. Ibid.
4. Ibid.
5. A. Chevallier, *The Encyclopedia of Medicinal Plants*, 1st American ed. (New York: DK, 1996), 98.
6. A. Plaitakis and R. C. Duvoisin, "Homer's Moly Identified as *Galanthus nivalis* L.: Physiologic Antidote to Stramonium Poisoning," *Clinical Neuropharmacology* 6, no. 1 (1983): 1–5.
7. E. Freye, *Pharmacology and Abuse of Cocaine, Amphetamines, Ecstasy, and Related Designer Drugs* (New York: Springer, 2009), 217–18.
8. U. Quattrocchi, *CRC World Dictionary of Medicinal and Poisonous Plants: Common Names, Scientific Names, Eponyms, Synonyms, and Etymology* (Boca Raton, FL: CRC, 2012), 637.
9. Chevallier, *The Encyclopedia of Medicinal Plants*.
10. http://www.theoi.com/Gallery/T35.6.html.
11. Hoffmann, *Modern Magic: A Practical Treatise on the Art of Conjuring*, 9th ed. (London: G. Routledge and Sons, 1894).
12. M. D. Mashkovsky and R. P. Kruglikova-Lvova, "On the Pharmacology of the New Alkaloid Galantamine," *Farmakologia Toxicologia* 14 (1951): 27–30.
13. Ibid.
14. Ibid.
15. Ibid.
16. P. White, et al., "Neocortical Cholinergic Neurons in Elderly People," *Lancet* 1, no. 8013 (March 26, 1977): 668–71.
17. Ibid.
18. Plaitakis and Duvoisin, "Homer's Moly Identified as *Galanthus nivalis* L."
19. Ibid.
20. Ibid.
21. Theophrastus, *Historia plantarum* (Treviso, Italy: Bartholomaeus Confalonerius, 1483), leaves.
22. M. Vaughn, *Stardust* (Paramount Pictures, 2007).
23. E. Crundwell, "The Unnatural History of the Fly Agaric," *Mycologist* 1, no. 4 (1987): 178–81.
24. N. Price, *The Viking Way: Religion and War in Iron Age Scandanavia*, ed. O. Kyhlberg (Uppsala, Sweden: Uppsala University, 2002), 366.
25. V. P. Wasson and R. G. Wasson, *Mushrooms, Russia, and History* (New York: Pantheon, 1957), 192–93.
26. Ibid.
27. Crundwell, "The Unnatural History of the Fly Agaric"; R. Wasson, *Soma: Divine Mushroom of Immortality* (New York: Harcourt Brace, 1968), 348–55; D. Michelot and L. M. Melendez-Howell, "*Amanita muscaria*: Chemistry, Biology, Toxicology, and Ethnomycology," *Mycological Research* 107, no. 2 (2003): 131–46; S. Hajicek-Dobberstein, "Soma Siddhas and Alchemical Enlightenment: Psychedelic Mushrooms in Buddhist Tradition," *Journal of Ethnopharmacology* 48, no. 2 (1995): 99–118; S. Wolfton, *The Fly Agaric: Berserker's Boon or Shaman's Elixir*

(Reading, UK: Coxland Press, 1994); and C. Li and N. H. Oberlies, "The Most Widely Recognized Mushroom: Chemistry of the Genus *Amanita*," *Life Sciences* 78, no. 5 (2005): 532–38.

28. R. B. Anderson, *Norse Mythology or the Religion of Our Forefathers: Containing All the Myths of the Eddas* (Chicago: 1879).

29. R. Keyser and B. Pennock, *The Religion of the Northmen* (New York: C. B. Norton, 1854).

30. Hajicek-Dobberstein, "Soma Siddhas and Alchemical Enlightenment."

31. Ibid.

32. Michelot and Melendez-Howell, "*Amanita muscaria*."

33. L. Millman, *Giant Polypores and Stoned Reindeer* (Cambridge, MA: Komatik Press, 2013), 49–53.

34. Wasson, *Mushrooms, Russia, and History*, 194–95.

35. Crundwell, "The Unnatural History of the Fly Agaric"; and Price, *The Viking Way*.

36. Price, *The Viking Way*.

37. Quattrocchi, *CRC World Dictionary of Medicinal and Poisonous Plants*, 530.

38. Chevallier, *The Encyclopedia of Medicinal Plants*, 219.

39. Ibid.

40. Quattrocchi, *CRC World Dictionary of Medicinal and Poisonous Plants*, 530.

41. Chevallier, *The Encyclopedia of Medicinal Plants*, 219.

42. Ibid.

43. Price, *The Viking Way*.

44. G. M. Smith and H. K. Beecher, "Amphetamine Sulfate and Athletic Performance. I. Objective Effects," *Journal of the American Medical Association* 170, no. 5 (1959): 542–57.

45. F. Sjöqvist, M. Garle, and A. Rane, "Use of Doping Agents, Particularly Anabolic Steroids, in Sports and Society," *Lancet* 371, no. 9627 (2008): 1872–82.

46. R. Mukhopadhyay, "Catching Doping Athletes," *Analytical Chemistry* 79, no. 15 (2007): 5522–28.

47. Ibid.

48. D. G. Liddle and D. J. Connor, "Nutritional Supplements and Ergogenic Aids," *Primary Care* 40, no. 2 (2013): 487–505.

49. P. Hemmersbach, "History of Mass Spectrometry at the Olympic Games," *Journal of Mass Spectrometry* 43, no. 7 (2008): 839–53.

50. Liddle and Connor, "Nutritional Supplements and Ergogenic Aids."

51. "Gene Doping," *Play True* 1 (2005): 1–9.

52. R. Friedman, "Why Are We Drugging Our Soldiers?" *New York Times*, April 21, 2012.

53. Ibid.

54. Ibid.

55. Ibid.

56. Ibid.

57. Ovid, *Ovid's Metamorphoses, in Fifteen Books. Translated by Mr. Dryden. Mr. Addison. Dr. Garth. Mr. Mainwaring. Mr. Congreve. Mr. Rowe. Mr. Pope. Mr. Gay. Mr. Eusden. Mr. Croxall. And other eminent hands. Published by Sir Samuel Garth, M.D. Adorn'd with sculptures. Volume the first* (London: Printed for J. and R. Tonson and S. Draper in the Strand, 1751), bk. 4 (Hermaphroditus).

58. Ibid.

59. W. C. Hembree et al., "Endocrine Treatment of Transsexual Persons: An Endocrine Society Clinical Practice Guideline," *Journal of Clinical Endocrinology and Metabolism* 94, no. 9 (2009): 3132–54.
60. Ibid.
61. Ibid.
62. Ibid.

## CHAPTER THREE: IMMORTALITY AND LONGEVITY

1. J. C. Cooper, *Chinese Alchemy: The Taoist Quest for Immortality* (Wellingborough, UK: Aquarian Press, 1984), 13–28 and 44–45.
2. Ibid.
3. Ibid.
4. Ibid.
5. Ibid.
6. Ibid.
7. Ibid.
8. C. Finch, *Longevity, Senescence, and the Genome*, John D. and Catherine T. MacArthur Foundation Series on Mental Health and Development (Chicago: University of Chicago Press, 1990).
9. J. D. Congdon et al., "Hypotheses of Aging in a Long-Lived Vertebrate, Blanding's Turtle (*Emydoidea blandingii*)," *Experimental Gerontology* 36, nos. 4–6 (2001): 813–27; and J. D. Congdon et al., "Testing Hypotheses of Aging in Long-Lived Painted Turtles (*Chrysemys picta*)," *Experimental Gerontology* 38, no. 7 (2003): 765–72.
10. G. R. Acsádi and J. N. Nemeskéri, *History of Human Life Span and Mortality* (Budapest: Akadémiai Kiadó, 1970).
11. P. L. Lutz, H. M. Prentice, and S. L. Milton, "Is Turtle Longevity Linked to Enhanced Mechanisms for Surviving Brain Anoxia and Reoxygenation?" *Experimental Gerontology* 38, no. 7 (2003): 797–800.
12. Ibid.
13. Ibid.
14. M. E. Rice, E. J. Lee, and Y. Choy, "High Levels of Ascorbic Acid, Not Glutathione, in the CNS of Anoxia-Tolerant Reptiles Contrasted with Levels in Anoxia-Intolerant Species," *Journal of Neurochemistry* 64, no. 4 (1995): 1790–99.
15. W. G. Willmore and K. B. Storey, "Glutathione Systems and Anoxia Tolerance in Turtles," *American Journal of Physiology* 273, no. 1 (1997): R219–25.
16. S. J. Olshansky and B. A. Carnes, *The Quest for Immortality: Science at the Frontiers of Aging* (New York: Norton, 2001), 139–42.
17. D. J. Grdina et al., "Protective Effect of S-2-(3-aminopropylamino)-ethylphosphorothioic Acid against Induction of Altered Hepatocyte Foci in Rats Treated Once with Gamma Radiation within One Day after Birth," *Cancer Research* 45, no. 11 (1985): 5379–81.
18. Olshansky and Carnes, *The Quest for Immortality*, 139–42; and D. J. Grdina et al., "The Radioprotector WR1065 Reduces Radiation-Induced Mutations at the Hypoxanthine-Guanine Phosphoribosyl Transferase Locus in V79 Cells," *Carcinogenesis* 6, no. 6 (1985): 929–31.
19. Olshansky and Carnes, *The Quest for Immortality*, 139–42.
20. Cooper, *Chinese Alchemy*.
21. Ibid.
22. Ibid.; Olshansky and Carnes, *The Quest for Immortality*; and D. C. Wright, *The*

*History of China*, Greenwood Histories of the Modern Nations (Westport, CT: Greenwood Press, 2001).

23. Olshansky and Carnes, *The Quest for Immortality*, 46.
24. Ibid., 29.
25. Ibid., 32; and Cooper, *Chinese Alchemy*.
26. Olshansky and Carnes, *The Quest for Immortality*, 41.
27. J. K. Rowling and M. GrandPré, *Harry Potter and the Sorcerer's Stone*, 1st American ed. (New York: A. A. Levine Books, 1998).
28. J. Emsley, *The Shocking History of Phosphorus: A Biography of the Devil's Element* (London: Macmillan, 2000).
29. K. Ashley, D. Cordell, and D. Mavinic, "A Brief History of Phosphorus: From the Philosopher's Stone to Nutrient Recovery and Reuse," *Chemosphere* 84, no. 6 (2011): 737–46.
30. Emsley, *The Shocking History of Phosphorus*.
31. Ibid.
32. Ashley, Cordell, and Mavinic, "A Brief History of Phosphorus."
33. Ibid.
34. Ibid.
35. Ibid.
36. A. Mayor, *The Poison King: The Life and Legend of Mithradates, Rome's Deadliest Enemy* (Princeton, NJ: Princeton University Press, 2010), 240–47.
37. Ibid.
38. Ibid.
39. Ibid.
40. Ibid.
41. R. Reiner, *The Princess Bride* (Warner Bros., 1987).
42. Mayor, *The Poison King*.
43. L. B. Moore et al., "St. John's Wort Induces Hepatic Drug Metabolism through Activation of the Pregnane X Receptor," *Proceedings of the National Academy of Sciences USA* 97, no. 13 (2000): 7500–7502.
44. J. C. Whorton, *The Arsenic Century: How Victorian Britain Was Poisoned at Home, Work, and Play* (Oxford, UK: Oxford University Press, 2010), 10.
45. Ibid., 18.
46. Ibid., 19.
47. R. J. Flanagan and A. L. Jones, *Antidotes* (New York: Taylor & Francis, 2001), 11–13 and 172–73.
48. Ibid.; and E. Helftenstein, "Lorenzo de' Medici's Magnificent Cups: Precious Vessels as Status Symbols in Fifteenth Century Europe," *I Tatti Studies in the Italian Renaissance* 16, no. ½ (2013): 415–44.
49. G. Zammit-Maempel, *Pioneers of Maltese Geology*, 1st ed. (Malta: Mid-Med Bank, 1989).
50. G. Zammit-Maempel, "The Folklore of Maltese Fossils," *Papers in Mediterranean Social Studies* 1, no. 1 (1990): 20–21.
51. Zammit-Maempel, *Pioneers of Maltese Geology*.
52. H. C. Kee, *The Cambridge Annotated Study Bible: New Revised Standard Version* (Cambridge, UK: Cambridge University Press, 1993), 4.
53. Ibid., 7.
54. Olshansky and Carnes, *The Quest for Immortality*, 85.
55. Ibid.

56. J. Gearhart, "New Potential for Human Embryonic Stem Cells," *Science* 282, no. 5391 (1998): 1061–62.
57. Olshansky and Carnes, *The Quest for Immortality*, 130–33.
58. Ibid.
59. Ibid.
60. K. Takahashi and S. Yamanaka, "Induction of Pluripotent Stem Cells from Mouse Embryonic and Adult Fibroblast Cultures by Defined Factors," *Cell* 126, no. 4 (2006): 663–76.
61. D. E. Harrison et al., "Rapamycin Fed Late in Life Extends Lifespan in Genetically Heterogeneous Mice," *Nature* 460, no. 7253 (2009): 392–95.
62. Ibid.
63. L. Fontana, L. Partridge, and V. D. Longo, "Extending Healthy Life Span—from Yeast to Humans," *Science* 328, no. 5976 (2010): 321–26.
64. L. Fontana et al., "Long-Term Effects of Calorie or Protein Restriction on Serum IGF-1 and IGFBP-3 Concentration in Humans," *Aging Cell* 7, no. 5 (2008): 681–87.
65. Ibid.
66. V. D. Longo and M. P. Mattson, "Fasting: Molecular Mechanisms and Clinical Applications," *Cell Metabolism* 19, no. 2 (2014): 181–92.

**CHAPTER FOUR: SUPERNATURAL SKIES**
1. C. Renfrew and P. G. Bahn, *Archaeology: Theories, Methods, and Practice*, 3rd ed. (New York: Thames and Hudson, 2000), 197.
2. Ibid.
3. D. Jones, "New Light on Stonehenge," *Smithsonian*, October 2008.
4. M. Pitts, "Gerald Hawkins: Astronomer Who Claimed Stonehenge Was a Computer," *Guardian*, July 24, 2003.
5. M. Parker Pearson and Stonehenge Riverside Project (England), *Stonehenge: A New Understanding; Solving the Mysteries of the Greatest Stone Age Monument* (New York: The Experiment, 2013).
6. J. Wozencroft and P. Devereux, "Stone Age Eyes and Ears: A Visual and Acoustic Pilot Study of Carn Menyn and Environs, Preseli, Wales," *Time and Mind: The Journal of Archaeology, Consciousness, and Culture* 7, no. 1 (2014): 47–70.
7. R. Keyser and B. Pennock, *The Religion of the Northmen* (New York: C. B. Norton, 1854).
8. V. Ions, *Egyptian Mythology* (New York: Peter Bedrick Books, 1982).
9. http://eclipse.gsfc.nasa.gov/.
10. Ibid.
11. C. Cajochen et al., "Evidence That the Lunar Cycle Influences Human Sleep," *Current Biology* 23, no. 15 (2013): 1485–88.
12. M. Wikelski and M. Hau, "Is There an Endogenous Tidal Foraging Rhythm in Marine Iguanas?" *Journal of Biological Rhythms* 10, no. 4 (1995): 335–50.
13. Cajochen et al., "Evidence That the Lunar Cycle Influences Human Sleep."
14. M. Smith, I. Croy, and K. Persson Waye, "Human Sleep and Cortical Reactivity Are Influenced by Lunar Phase," *Current Biology* 24, no. 12 (2014): R551–52; and C. Z. Turanyi et al., "Association between Lunar Phase and Sleep Characteristics," *Sleep Medicine* 15, no. 11 (2014): 1411–16.
15. M. Cordi et al., "Lunar Cycle Effects on Sleep and the File Drawer Problem," *Current Biology* 24, no. 12 (2014): R549–50.

16. C. W. Marean et al., "Early Human Use of Marine Resources and Pigment in South Africa during the Middle Pleistocene," *Nature* 449, no. 7164 (2007): 905–8.

17. S. Thirslund, *Viking Navigation* (Roskilde, Denmark: Viking Ship Museum, 2007); and G. P. Helgadóttir, ed., *Hrafns saga Sveinbjarnarsonar* (Oxford, UK: Oxford University Press; New York: Clarendon Press, 1987).

18. J. Haywood, *The Penguin Historical Atlas of the Vikings* (London: Penguin, 1995), 32–33.

19. Thirslund, *Viking Navigation*.

20. J. Stetson, "Viking Sun Stones," http://www.atoptics.co.uk/fz767.htm.

21. Thirslund, *Viking Navigation*.

22. A. Le Floch et al., "The Sixteenth Century Alderney Crystal: A Calcite as an Efficient Reference Optical Compass?" *Proceedings of the Royal Society A: Mathematical, Physical, and Engineering Sciences* 469, no. 2153 (2013): 651.

23. G. Ropars et al., "A Depolarizer as a Possible Precise Sunstone for Viking Navigation by Polarized Skylight," *Proceedings of the Royal Society A: Mathematical, Physical and Engineering Sciences* 468, no. 2139 (2011): 671–84.

24. Ibid.

25. C. Roslund and C. Beckman, "Disputing Viking Navigation by Polarized Skylight," *Applied Optics* 33, no. 21 (1994): 4754–55.

26. H. C. Kee, *The Cambridge Annotated Study Bible: New Revised Standard Version* (Cambridge, UK: Cambridge University Press, 1993), 54.

27. B. J. Sivertsen, *The Parting of the Sea: How Volcanoes, Earthquakes, and Plagues Shaped the Story of Exodus* (Princeton, NJ: Princeton University Press, 2009), xix, 35, and 239.

28. *Amazing Fantasy* (New York: Atlas Magazines, 1962), no. 1; and M. Webb, *The Amazing Spider-Man* (Columbia Pictures, 2012).

29. A. B. Lord, *The Singer of Tales*, Harvard Studies in Comparative Literature (Cambridge, MA: Harvard University Press, 1960).

30. Ibid.

31. Ibid.

32. J. Garstang and J. Garstang, *The Story of Jericho*, new ed. (London: Marshall, 1948).

33. Ibid.

34. Ibid.; and Sivertsen, *The Parting of the Sea*, 7.

35. B. H. Baker, P. A. Mohr, and L. A. J. Williams, *Geology of the Eastern Rift System of Africa*, Geological Society of America Special Paper 136 (Boulder, CO: Geological Society of America, 1972).

36. G. Hort, "The Plagues of Egypt," *Zeitschrift für die alttestamentliche Wissenschaft* 69, nos. 1–4 (1957): 84–103.

37. Sivertsen, *The Parting of the Sea*, 7.

38. B. G. Mason, D. M. Pyle, and C. Oppenheimer, "The Size and Frequency of the Largest Explosive Eruptions on Earth," *Bulletin of Volcanology* 66, no. 8 (2004): 735–48.

39. C. G. Newhall and S. Self, "The Volcanic Explosivity Index (VEI): An Estimate of Explosive Magnitude for Historical Volcanism," *Journal of Geophysical Research: Oceans* 87, no. C2 (1982): 1231–38.

40. Sivertsen, *The Parting of the Sea*, 37–42.

41. Kee, *The Cambridge Annotated Study Bible*, 52.

42. Sivertsen, *The Parting of the Sea*, 37–42.

43. R. Schoental, "Mycotoxins and the Bible," *Perspectives in Biological Medicine* 28, no. 1 (1984): 117–20.
44. Sivertsen, *The Parting of the Sea*, 37–42.
45. Kee, *The Cambridge Annotated Study Bible*, 52.
46. Sivertsen, *The Parting of the Sea*.
47. J. S. Marr and C. D. Malloy, "An Epidemiologic Analysis of the Ten Plagues of Egypt," *Caduceus* 12, no. 1 (1996): 7–24.
48. B. M. Vinther et al., "A Synchronized Dating of Three Greenland Ice Cores throughout the Holocen," *Journal of Geophysical Research: Atmosphere* 111, no. D13 (2006).
49. E. H. Cline, *From Eden to Exile: Unraveling Mysteries of the Bible* (Washington, DC: National Geographic, 2007), 71–73.
50. N. J. G. Pearce, J. A. Westgate, S. J. Preece, W. J. Eastwood, and W. T. Perkins, "Identification of Aniakchak (Alaska) Tephra in Greenland Ice Core Challenges the 1645 BC Date for Minoan Eruption of Santorini," *Geochemistry Geophysics Geosystems* 5, no. 3 (March 2004).
51. Sivertsen, *The Parting of the Sea*, 44–45.
52. D. B. Vitaliano, "Geomythology: Geological Origins of Myths and Legends," *Geological Society, London, Special Publications* 273, no. 1 (2007): 1–7; and D. B. Vitaliano, *Legends of the Earth: Their Geologic Origins* (Bloomington: Indiana University Press, 1973).
53. Ibid.
54. Ibid.
55. Sivertsen, *The Parting of the Sea*.
56. R. Zemeckis, *Back to the Future* (Universal Pictures, 1985).
57. J. Kasparian et al., "Electric Events Synchronized with Laser Filaments in Thunderclouds," *Optics Express* 16, no. 8 (2008): 5757–63.
58. Ibid.
59. "It Never Rains," *Economist*, December 13, 2006, http://www.economist.com/node/8413323; "Clouds over Troubled Waters," *Economist*, September 20, 2012, http://www.economist.com/node/21563279.
60. Ibid.
61. Ibid.
62. Ibid.
63. Ibid.
64. J. Latham et al., "Weakening of Hurricanes via Marine Cloud Brightening (MCB)," *Atmospheric Science Letters* 13, no. 4 (2012): 231–37.
65. Ibid.
66. E. Blake, C. Landsea, and E. Gibney, *The Deadliest, Costliest, and Most Intense United States Tropical Cyclones from 1851 to 2010* (National Hurricane Center, 2011).
67. Latham et al., "Weakening of Hurricanes via Marine Cloud Brightening (MCB)."

## CHAPTER FIVE: ANIMALS AND PLANTS AS OMENS, GUIDES, AND GODS

1. R. B. Anderson, *Norse Mythology or the Religion of Our Forefathers: Containing All the Myths of the Eddas* (Chicago: 1879), 219–21.
2. S. Hajicek-Dobberstein, "Soma Siddhas and Alchemical Enlightenment: Psychedelic Mushrooms in Buddhist Tradition," *Journal of Ethnopharmacology* 48, no. 2 (1995): 99–118.

3. Anderson, *Norse Mythology or the Religion of Our Forefathers*, 219–21.

4. D. Stahler, B. Heinrich, and D. Smith, "Common Ravens, *Corvus corax*, Preferentially Associate with Grey Wolves, *Canis lupus*, as a Foraging Strategy in Winter," *Animal Behaviour* 64, no. 2 (2002): 283–90.

5. D. Stahler, "Interspecific Interactions between the Common Raven (*Corvus corax*) and the Gray Wolf (*Canis lupus*) in Yellowstone National Park, Wyoming: Investigations of a Predator and Scavenger Relationship," master's thesis, University of Vermont, 2000.

6. B. Thorpe et al., ed., *The Elder Eddas of Saemund Sigfusson*, Norrœna, the History and Romance of Northern Europe, Viking edition (London: Norrœna Society, 1906).

7. H. T. Bunn, "Hunting, Power Scavenging, and Butchering by Hadza Foragers and by Plio-Pleistocene Homo," in *Meat-Eating and Human Evolution*, ed. C. B. Stanford and H. T. Bunn (Oxford, UK: Oxford University Press, 2001), 199–218.

8. H. C. Kee, *The Cambridge Annotated Study Bible: New Revised Standard Version* (Cambridge, UK: Cambridge University Press, 1993).

9. H. Waddell, *The Desert Fathers: Translations from the Latin*, 1st ed., Vintage Spiritual Classics (New York: Vintage Books, 1998).

10. Ibid.

11. Stahler, "Interspecific Interactions between the Common Raven (*Corvus corax*) and the Gray Wolf (*Canis lupus*) in Yellowstone National Park, Wyoming."

12. K. Kishimoto et al., "Volatile C6-Aldehydes and Allo-Ocimene Activate Defense Genes and Induce Resistance against *Botrytis cinerea* in *Arabidopsis thaliana*," *Plant & Cell Physiology* 46, no. 7 (2005): 1093–102.

13. S. Gomez et al., "Costs and Benefits of Induced Resistance in a Clonal Plant Network," *Oecologia* 153, no. 4 (2007): 921–30.

14. Z. Babikova et al., "Underground Signals Carried through Common Mycelial Networks Warn Neighbouring Plants of Aphid Attack," *Ecology Letters* 16, no. 7 (2013): 835–43.

15. Y. Y. Song, R. S. Zeng, J. F. Xu, J. Li, X. Shen, and W. G. Yihdego, "Interplant Communication of Tomato Plants through Underground Common Mycorrhizal Networks," *PLoS One* 5, no. 10 (October 13, 2010): e13324.

16. Ibid.

17. Ibid.

18. J. Cameron, *Avatar* (Twentieth Century Fox, 2009).

19. D. M. Dosa, "A Day in the Life of Oscar the Cat," *New England Journal of Medicine* 357, no. 4 (2007): 328–29.

20. R. Ehmann et al., "Canine Scent Detection in the Diagnosis of Lung Cancer: Revisiting a Puzzling Phenomenon," *European Respiratory Journal* 39, no. 3 (2012): 669–76.

21. J. N. Cornu et al., "Olfactory Detection of Prostate Cancer by Dogs Sniffing Urine: A Step Forward in Early Diagnosis," *European Urology* 59, no. 2 (2011): 197–201.

22. M. K. Bomers et al., "Using a Dog's Superior Olfactory Sensitivity to Identify *Clostridium difficile* in Stools and Patients: Proof of Principle Study," *BMJ* 345 (2012): e7396.

23. C. W. Breuner et al., "Environment, Behavior, and Physiology: Do Birds Use Barometric Pressure to Predict Storms?" *Journal of Experimental Biology* 216, no. 11 (2013): 1982–90.

24. H. M. Streby et al., "Tornadic Storm Avoidance Behavior in Breeding Songbirds," *Current Biology* 25, no. 1 (2015): 98–102.

25. Aelian, *On the Nature of Animals*, ed. G. McNamee (San Antonio, TX: Trinity University Press, 2011).

26. M. Kaplan, "When Animals Predict Earthquakes," *New Scientist*, no. 2591 (February 17, 2007).

27. Ibid.

28. S. Coren, "Can Dogs Predict Earthquakes?" paper presented at the 47th Annual Meeting of the Psychonomic Society, November 16–19, 2006, Houston, Texas.

29. Ibid.

30. Ibid.

31. Kaplan, "When Animals Predict Earthquakes."

32. J. T. Hagstrum, "Atmospheric Propagation Modeling Indicates Homing Pigeons Use Loft-Specific Infrasonic 'Map' Cues," *Journal of Experimental Biology* 216, no. 4 (2013): 687–99.

33. J. Grimm and W. Grimm, *The Fairy Ring: A New Collection of Popular Tales*, trans. J. E. Taylor (London: A. Hart, 1857), 86–88.

34. M. Tatar, *The Classic Fairy Tales: Texts, Criticism*, 1st Norton Critical ed. (New York: Norton, 1999).

35. Ibid.

36. Ibid.

37. Grimm and Grimm, *The Fairy Ring*.

38. Tatar, *The Classic Fairy Tales*.

39. Livy and W. M. Roberts, *The History of Rome* (London: J. M. Dent & Sons; New York: E. P. Dutton, 1921).

40. C. Dio, *Dio's Roman History*, trans. E. Cary and H. B. Foster, Loeb Classical Library (London: W. Heinemann; New York: Macmillan, 1914).

41. Homer, *The Iliad*, trans. S. Butler (St. Petersburg, FL: Red and Black, 2008).

42. D. Collins, "Reading the Birds: Oionomanteia in Early Epic," *Colby Quarterly* 38, no. 1 (2002).

43. Ibid.

44. Ibid.

45. Ibid.

46. Ibid.

47. "Frequently Asked Questions about El Niño and La Niña," http://www.cpc.noaa.gov/products/analysis_monitoring/ensostuff/ensofaq.shtml#DIFFER.

48. J. Shaman and M. Lipsitch, "The El Niño–Southern Oscillation (ENSO)–Pandemic Influenza Connection: Coincident or Causal?" *Proceedings of the National Academy of Sciences USA* 110, supplement 1 (2013): 3689–91.

49. M. Pascual et al., "Cholera Dynamics and El Niño–Southern Oscillation," *Science* 289, no. 5485 (2000): 1766–69; M. A. Johansson, D. A. Cummings, and G. E. Glass, "Multiyear Climate Variability and Dengue–El Niño Southern Oscillation, Weather, and Dengue Incidence in Puerto Rico, Mexico, and Thailand: A Longitudinal Data Analysis," *PLoS Medicine* 6, no. 11 (2009): e1000168; and M. J. Bouma and C. Dye, "Cycles of Malaria Associated with El Niño in Venezuela," *Journal of the American Medical Association* 278, no. 21 (1997): 1772–74.

50. "Frequently Asked Questions about El Niño and La Niña"; and S. Ineson and A. A. Scaife, "The Role of the Stratosphere in the European Climate Response to El Niño," *Nature Geoscience* 2, no. 1 (2008): 32–36.

51. J. Shamoun-Baranes et al., "Is There a Connection between Weather at Departure Sites, Onset of Migration, and Timing of Soaring-Bird Autumn Migration in Israel?" *Global Ecology and Biogeography* 15 (2006): 541–52; and "Bird Migration Routes," Kruger Park Birding, http://birding.krugerpark.co.za/birding-in-kruger-migration-routes.html.

52. U. Mellone et al., "Extremely Detoured Migration in an Inexperienced Bird: Interplay of Transport Costs and Social Interactions," *Journal of Avian Biology* 42, no. 5 (2011): 468–72.

53. S. Brönnimann et al., "ENSO Influence on Europe during the Last Centuries," *Climate Dynamics* 28 (2007): 181–97.

## CHAPTER SIX: PROPHECY

1. Hesiod and Homer, *Hesiod, the Homeric Hymns, and Homerica*, trans. H. G. Evelyn-White, Loeb Classical Library (London: W. Heinemann, 1914; New York: Macmillan, 1914), lines 507–44.

2. Strabo, *The Geography of Strabo* (Cambridge, MA: Harvard University Press, 1924), bk. 13, chap. 7, fragment 1c.

3. S. J. Browne and N. J. Aebischer, "Temporal Changes in the Migration Phenology of Turtle Doves *Streptopelia turtur* in Britain, Based on Sightings from Coastal Bird Observatories," *Journal of Avian Biology* 34, no. 1 (2003): 65–71; C. Eraud et al., "Migration Routes and Staging Areas of Trans-Saharan Turtle Doves Appraised from Light-Level Geolocators," *PloS One* 8, no. 3 (2013); P. Tryjanowski, S. Kuzniak, and T. H. Sparks, "Earlier Arrival of Some Farmland Migrants in Western Poland," *Ibis* 144, no. 1 (2002): 62–68; and P. Tryjanowski, S. Kuzniak, and T. H. Sparks, "What Affects the Magnitude of Change in First Arrival Dates of Migrant Birds?" *Journal of Ornithology* 146, no. 3 (2005): 200–205.

4. R. Graves, *Greek Myths*, illustrated ed. (London: Penguin Books, 1984), 178.

5. T. Curnow, *The Oracles of the Ancient World: A Complete Guide* (London: Duckworth, 2004), 20–23.

6. Ibid.

7. Ibid.

8. A. F. Aveni, *Behind the Crystal Ball: Magic, Science, and the Occult from Antiquity through the New Age*, 1st ed. (New York: Times Books, 1996), 22–23.

9. D. Collins, "Mapping the Entrails: The Practice of Greek Hepatoscopy," *American Journal of Philology* 129, no. 3 (2008): 319–45.

10. Ibid.

11. Ibid.

12. Ibid.

13. Ibid.

14. Ibid.

15. Aveni, *Behind the Crystal Ball*.

16. Collins, "Mapping the Entrails."

17. W. J. Broad, *The Oracle: The Lost Secrets and Hidden Message of Ancient Delphi* (New York: Penguin Press, 2006).

18. Ibid.

19. Plutarch, *Plutarch's Lives* (Cambridge, MA: Harvard University Press, 1914), chap. 14.

20. Broad, *The Oracle*.

21. J. A. S. Evans, "The Oracle of the Wooden Wall," *Classical Journal* 78, no. 1 (1982): 24–29.
22. Ibid.
23. Broad, *The Oracle*.
24. Ibid.
25. Ibid.
26. J. E. Fontenrose, *The Delphic Oracle: Its Responses and Operations, with a Catalogue of Responses* (Berkeley: University of California Press, 1978).
27. Diodorus, *Siculus: Historical Library*, Loeb Classical Libraries, vol. 289 (Cambridge, MA: Harvard University Press, 1952).
28. Ibid.
29. Ibid.
30. J. R. Hale et al., "Questioning the Delphic Oracle," *Scientific American* 289, no. 2 (2003): 66–73.
31. Ibid.
32. J. Z. de Boer, J. R. Hale, and J. Chanton, "New Evidence for the Geological Origins of the Ancient Delphic Oracle (Greece)," *Geology* 29, no. 8 (2001): 707.
33. Hale et al., "Questioning the Delphic Oracle."
34. Ibid.
35. Ibid.
36. Ibid.
37. De Boer, Hale, and Chanton, "New Evidence for the Geological Origins of the Ancient Delphic Oracle (Greece)."
38. Ibid.
39. Ibid.
40. Ibid.
41. A. Belcher and W. Sinnott-Armstrong, "Neurolaw," *Wiley Interdisciplinary Reviews: Cognitive Science* 1, no. 1 (2010): 18–22.
42. D. D. Langleben et al., "Brain Activity During Simulated Deception: An Event-Related Functional Magnetic Resonance Study," *Neuroimage* 15, no. 3 (2002): 727–32.
43. Ibid.
44. S. V. Shinkareva et al., "Using fMRI Brain Activation to Identify Cognitive States Associated with Perception of Tools and Dwellings," *PLoS One* 3, no. 1 (2008): e1394.
45. B. N. Pasley et al., "Reconstructing Speech from Human Auditory Cortex," *PLoS Biology* 10, no. 1 (2012): e1001251.
46. Ibid.

## CHAPTER SEVEN: BEYOND THE GRAVE

1. D. L. Paxson, *The Way of the Oracle: Recovering the Practices of the Past to Find Answers for Today* (San Francisco, CA: Weiser Books, 2012), 21.
2. R. Stillwell, W. L. MacDonald, and M. H. McAllister, *The Princeton Encyclopedia of Classical Sites* (Princeton, NJ: Princeton University Press, 1976), 251.
3. Ibid.
4. S. Caliro et al., "Geochemical and Biochemical Evidence of Lake Overturn and Fish Kill at Lake Averno, Italy," *Journal of Volcanology and Geothermal Research* 178, no. 2 (2008): 305–16.
5. J. Cabassi, "Geochemical Features of Nutrients and Dissolved Gases in the Vol-

canic Lake of Averno (Phlegrean Fields, Southern Italy)," *Geophysical Research Abstracts* 14 (2012).

6.  Caliro et al., "Geochemical and Biochemical Evidence of Lake Overturn and Fish Kill at Lake Averno, Italy."

7.  Virgil, *The Aeneid*, trans. R. Fitzgerald and J. Dryden, Classics of Greece and Rome Series (New York: Macmillan, 1965), chap. 6, lines 236–60.

8.  Stillwell, MacDonald, and McAllister, *The Princeton Encyclopedia of Classical Sites*, 391.

9.  Strabo, *The Geography of Strabo* (Cambridge, MA: Harvard University Press, 1924), bk. 13, chap. 4, sec. 14.

10.  Ibid.

11.  H. Pfanz et al., "The Ancient Gates to Hell and Their Relevance to Geogenic $CO_2$," in *History of Toxicology and Environmental Health*, ed. P. Wexler (Waltham, MA: Academic Press, 2014), 92–113.

12.  Ibid.

13.  C. Lambertsen, "Therapeutic Gases—Oxygen, Carbon Dioxide, and Helium," in *Drill's Pharmacology in Medicine*, ed. J. DiPalma (New York: McGraw-Hill, 1971).

14.  Ibid.

15.  Pfanz et al., "The Ancient Gates to Hell and Their Relevance to Geogenic $CO_2$."

16.  Strabo, *The Geography of Strabo*, bk. 13, chap. 4, sec. 14.

17.  Ibid.

18.  Ibid.

19.  J. Soentgen, "On the History and Prehistory of $CO_2$," *Foundations of Chemistry* 12, no. 2 (2009): 137–48.

20.  Pfanz et al., "The Ancient Gates to Hell and Their Relevance to Geogenic $CO_2$."

21.  R. J. Sharer and L. P. Traxler, *The Ancient Maya* (Stanford, CA: Stanford University Press, 2006), 565, 605, 752–53.

22.  Ibid.

23.  Ibid.

24.  Ibid.

25.  B. M. von Reumont et al., "The First Venomous Crustacean Revealed by Transcriptomics and Functional Morphology: Remipede Venom Glands Express a Unique Toxin Cocktail Dominated by Enzymes and a Neurotoxin," *Molecular Biology and Evolution* 31, no. 1 (2014): 48–58.

26.  Ibid.

27.  Ibid.

28.  Von Reumont et al., "The First Venomous Crustacean Revealed by Transcriptomics and Functional Morphology."

29.  J. L. van der Ham and B. E. Felgenhauer, "The Functional Morphology of the Putative Injecting Apparatus of *Speleonectes tanumekes* (Remipedia)," *Journal of Crustacean Biology* 27, no. 1 (2007): 1–9.

30.  Von Reumont et al., "The First Venomous Crustacean Revealed by Transcriptomics and Functional Morphology."

31.  Ibid.

32.  Ibid.

33.  Ibid.

34.  Sharer and Traxler, *The Ancient Maya*, 754.

35.  Ibid.

36.  Ibid.

37. S. A. Mitchell, "Odinn, Charms, and Mediumism: Hávamál 157 in Its Nordic and European Contexts," in *Old Norse Mythology in Comparative Perspective*, ed. P. Hermann, S. A. Mitchell, and J. P. Schjodt (Cambridge, MA: Harvard University Press, 2015).
38. Ibid.
39. Ibid.
40. Ibid.
41. D. Mobbs and C. Watt, "There Is Nothing Paranormal about Near-Death Experiences: How Neuroscience Can Explain Seeing Bright Lights, Meeting the Dead, or Being Convinced You Are One of Them," *Trends in Cognitive Sciences* 15, no. 10 (2011): 447–49.
42. Ibid.
43. Ibid.
44. Ibid.
45. O. Blanke and S. Arzy, "The Out-of-Body Experience: Disturbed Self-Processing at the Temporo-Parietal Junction," *Neuroscientist* 11, no. 1 (2005): 16–24.
46. J. Borjigin et al., "Surge of Neurophysiological Coherence and Connectivity in the Dying Brain," *Proceedings of the National Academy of Sciences USA* 110, no. 35 (2013): 14432–37.
47. P. van Lommel et al., "Near-Death Experience in Survivors of Cardiac Arrest: A Prospective Study in the Netherlands," *Lancet* 358, no. 9298 (2001): 2039–45.
48. Ibid.
49. Ibid.
50. Ibid.
51. Ibid.
52. Ibid.

## CHAPTER EIGHT: ENCHANTMENT

1. H. C. Kee, *The Cambridge Annotated Study Bible: New Revised Standard Version* (Cambridge, UK: Cambridge University Press, 1993), 2–3.
2. Ibid.
3. Ibid.
4. M. W. Johnson et al., "Pilot Study of the $5\text{-HT}_{2A}\text{R}$ Agonist Psilocybin in the Treatment of Tobacco Addiction," *Journal of Psychopharmacology* (September 11, 2014), doi: 10.1177/0269881114548296.
5. C. S. Grob et al., "Pilot Study of Psilocybin Treatment for Anxiety in Patients with Advanced-Stage Cancer," *Archives of General Psychiatry* 68, no. 1 (2011): 71–78.
6. Kee, *The Cambridge Annotated Study Bible*, 2–3.
7. Ibid.
8. Ibid.
9. D. Gillette, *The Shaman's Secret: The Lost Resurrection Teachings of the Ancient Maya* (New York: Bantam Books, 1997), 19–23.
10. Ibid.
11. Ibid.
12. Ibid., 50–57.
13. Ibid.
14. Ibid.
15. Ibid.

16. Ibid.
17. B. Bastian, J. Jetten, and M. J. Hornsey, "Gustatory Pleasure and Pain: The Offset of Acute Physical Pain Enhances Responsiveness to Taste," *Appetite* 72 (2014): 150–55.
18. Ibid.
19. Ibid.
20. Ibid.
21. Ibid.
22. Ibid.
23. Gillette, *The Shaman's Secret*, 50–57.
24. Ibid.
25. Apollodorus, *The Library*, trans. J. G. Frazer, Loeb Classical Library (London: W. Heinemann; New York: G. P. Putnam's Sons, 1921), bk. 1, chap. 9.
26. Ibid.
27. A. J. Carter, "Narcosis and Nightshade," *BMJ* 313, no. 7072 (1996): 1630–32.
28. Ibid.
29. Ibid.
30. Geoffrey of Monmouth, *History of the Kings of Britain*, trans. S. Evans, Everyman's Library (London: Dent, 1963).
31. W. Shakespeare, *The Works of Shakespeare*, ed. Alexander Pope (New York: AMS Press, 1969).
32. Ibid.
33. M. Tatar, *The Classic Fairy Tales: Texts, Criticism*, 1st Norton Critical ed. (New York: Norton, 1999).
34. T. Bulfinch, *Bulfinch's Mythology* (New York: Modern Library, 2004).
35. Apuleius, *The Apologia and Florida of Apuleius of Madaura*, trans. H. E. Butler (Oxford, UK: Clarendon Press, 1909).
36. Ibid.
37. R. Wagner, *Tristan and Isolde*, English National Opera Guide (London: J. Calder; New York: Riverrun Press, 1981).
38. Shakespeare, *The Works of Shakespeare*.
39. Ibid.
40. A. Chevallier, *The Encyclopedia of Medicinal Plants*, 1st American ed. (New York: DK, 1996).
41. Ibid.; and U. Quattrocchi, *CRC World Dictionary of Medicinal and Poisonous Plants: Common Names, Scientific Names, Eponyms, Synonyms, and Etymology* (Boca Raton, FL: CRC, 2012), 750.
42. Quattrocchi, *CRC World Dictionary of Medicinal and Poisonous Plants*, 750.
43. A. F. Aveni, *Behind the Crystal Ball: Magic, Science, and the Occult from Antiquity through the New Age*, 1st ed. (New York: Times Books, 1996), 136–37.
44. Ibid.
45. Ibid.
46. Ibid.
47. Ibid.
48. Ibid.
49. R. Mirza et al., "Do Marine Mollusks Possess Aphrodisiacal Properties?" *Abstracts of Papers of the American Chemical Society* 229 (2005).
50. G. D'Aniello et al., "D-Aspartate, a Key Element for the Improvement of Sperm Quality," *Advances in Sexual Medicine* 2, no. 4 (2012): 45–53.

51. P. Sandroni, "Aphrodisiacs Past and Present: A Historical Review," *Clinical Autonomic Research* 11, no. 5 (2001): 303–7.

52. Ibid.

53. Ibid.; and Chevallier, *The Encyclopedia of Medicinal Plants*, 116, 241, and 316.

54. B. Ditzen et al., "Intranasal Oxytocin Increases Positive Communication and Reduces Cortisol Levels during Couple Conflict," *Biological Psychiatry* 65, no. 9 (2009): 728–31.

55. G. Bedi, D. Hyman, and H. de Wit, "Is Ecstasy an 'Empathogen'? Effects of +/-3,4-methylenedioxymethamphetamine on Prosocial Feelings and Identification of Emotional States in Others," *Biological Psychiatry* 68, no. 12 (2010): 1134–40.

56. Ibid.

57. P. W. Eastwick and E. J. Finkel, "The Attachment System in Fledgling Relationships: An Activating Role for Attachment Anxiety," *Journal of Personality and Social Psychology* 95, no. 3 (2008): 628–47; P. W. Eastwick, S. D. Saigal, and E. J. Finkel, "Smooth Operating: A Structural Analysis of Social Behavior (SASB) Perspective on Initial Romantic Encounters," *Social Psychological and Personality Science* 1, no. 4 (2010): 344–52; E. J. Finkel and P. W. Eastwick, "Speed-Dating," *Current Directions in Psychological Science* 17, no. 3 (2008): 193–97; and E. J. Finkel, P. W. Eastwick, and J. Matthews, "Speed-Dating as an Invaluable Tool for Studying Romantic Attraction: A Methodological Primer," *Personal Relationships* 14, no. 1 (2007): 149–66.

58. H. E. Fisher, "Lust, Attraction, Attachment: Biology and Evolution of the Three Primary Emotion Systems for Mating, Reproduction, and Parenting," *Journal of Sex Education and Therapy* 25, no. 1 (2000): 96–103.

59. H. E. Fisher, "Lust, Attraction, and Attachment in Mammalian Reproduction," *Human Nature: An Interdisciplinary Biosocial Perspective* 9, no. 1 (1998): 23–52.

60. H. E. Fisher et al., "Reward, Addiction, and Emotion Regulation Systems Associated with Rejection in Love," *Journal of Neurophysiology* 104, no. 1 (2010): 51–60.

61. Ibid.

62. Ibid.

63. L. J. Young et al., "Increased Affiliative Response to Vasopressin in Mice Expressing the V-1a Receptor from a Monogamous Vole," *Nature* 400, no. 6746 (1999): 766–68.

64. J. Wingfield, "Hormone-Behavior Interactions and Mating Systems in Male and Female Birds," in *The Differences between the Sexes*, ed. R. V. Short and E. Balaban (New York: Cambridge University Press, 1994), 303–30.

65. A. Booth and J. M. Dabbs, "Testosterone and Men's Marriages," *Social Forces* 72, no. 2 (1993): 463–77.

66. C. Nolan, *Batman Begins* (Warner Bros., 2005).

67. A. Wachowski and L. Wachowski, *The Matrix* (Warner Bros., 1999).

## CHAPTER NINE: SUPERHUMANS

1. B. Levinson, *Rain Man* (United Artists, 1988).

2. D. Tammet, *Embracing the Wide Sky: A Tour across the Horizons of the Mind* (New York: Free Press, 2009).

3. S. Baron-Cohen et al., "Talent in Autism: Hyper-Systemizing, Hyper-Attention to Detail, and Sensory Hypersensitivity," *Philosophical Transactions of the Royal Society of London, Series B, Biological Sciences* 364, no. 1522 (2009): 1377–83.

4. Ibid.
5. S. L. Macknik, S. Martinez-Conde, and S. Blakeslee, *Sleights of Mind: What the Neuroscience of Magic Reveals about Our Everyday Deceptions*, 1st ed. (New York: Henry Holt, 2010).
6. Ibid.
7. Ibid.
8. A. Snyder, "Explaining and Inducing Savant Skills: Privileged Access to Lower-Level, Less-Processed Information," *Philosophical Transactions of the Royal Society of London, Series B, Biological Sciences* 364, no. 1522 (2009): 1399–405.
9. Ibid.
10. D. Keyes, *Flowers for Algernon* (New York: Harcourt, 1966).
11. J. C. Naviaux et al., "Reversal of Autism-like Behaviors and Metabolism in Adult Mice with Single-Dose Antipurinergic Therapy," *Translational Psychiatry* 4 (2014).
12. Ibid.
13. Ibid.
14. Ibid.
15. Ibid.
16. Ibid.
17. B. Ratner, *X-Men: The Last Stand* (Twentieth Century Fox, 2006).
18. G. Roberts, "How Can a Man Hold His Breath for 22 Minutes? The Amazing Feat of World Record Holder Tom Sietas Explained," *Mail Online*, June 10, 2012.
19. Ibid.
20. Ibid.
21. E. E. Solberg et al., "The Effect of Meditation on Shooting Performance," *British Journal of Sports Medicine* 30, no. 4 (1996): 342–46.
22. M. Faulks, "Demonstration of the Inner Fire or Tummo Meditation," 2012, https://www.youtube.com/watch?v=GiJvD2BPOhA.
23. H. Benson et al., "Body-Temperature Changes during the Practice of g Tum-mo Yoga," *Nature* 295, no. 5846 (1982): 234–36.
24. Ibid.
25. Ibid.
26. Ibid.
27. Ibid.
28. Ibid.
29. R. Jevning et al., "Redistribution of Blood Flow in Acute Hypometabolic Behavior," *American Journal of Physiology* 235, no. 1 (1978): R89–92.
30. B. Witcombe and D. Meyer, "Sword Swallowing and Its Side Effects," *BMJ* 333, no. 7582 (2006): 1285–87.
31. E. Huizinga, "On Esophagoscopy and Sword-Swallowing," *Annals of Otology, Rhinology, and Laryngology* 78, no. 1 (1969): 32–39.
32. Ibid.
33. Witcombe and Meyer, "Sword Swallowing and Its Side Effects."
34. J. Sternfield, *Firewalk: The Psychology of Physical Immunity* (Stockbridge, MA: Berkshire House, 1992), 66–85.
35. Ibid.
36. H. C. Kee, *The Cambridge Annotated Study Bible: New Revised Standard Version* (Cambridge, UK: Cambridge University Press, 1993).

37. Sternfield, *Firewalk*.

38. B. Leikind and W. J. McCarthy, "An Investigation of Firewalking," *Skeptical Inquirer* 10, no. 23 (1985): 29–34.

39. Ibid.

40. Ibid.

41. Ibid.

42. A. F. Aveni, *Behind the Crystal Ball: Magic, Science, and the Occult from Antiquity through the New Age*, 1st ed. (New York: Times Books, 1996), 231–37.

43. Ibid.

44. Ibid.

45. Ibid.

46. J. Steinmeyer, *Hiding the Elephant: How Magicians Invented the Impossible and Learned to Disappear* (New York: Carroll & Graf, 2003), 5–10.

47. Ibid.

48. Ibid.

49. Ibid.

50. Hoffmann, *Modern Magic: A Practical Treatise on the Art of Conjuring*, 9th ed. (London: G. Routledge and Sons, 1894), 74.

51. R. F. Burton et al., *The Arabian Nights* (New York: Blue Ribbon Books, 1932).

52. Herodotus, *The Histories*, ed. A. Godley (Cambridge. MA: Harvard University Press, 1920), bk. 3, chap. 41.

53. S. Thompson, *Motif-Index of Folk-Literature: A Classification of Narrative Elements in Folktales, Ballads, Myths, Fables, Mediaeval Romances, Exempla, Fabliaux, Jest-Books, and Local Legends*, rev. and enl. ed. (Bloomington: Indiana University Press, 1989), vol. 1.

54. Hoffmann, *Modern Magic*.

55. Steinmeyer, *Hiding the Elephant*.

56. Ibid.

57. A. D. Barnosky, *Biodiversity Response to Climate Change in the Middle Pleistocene: The Porcupine Cave Fauna from Colorado* (Berkeley: University of California Press, 2004).

58. *The Merriam-Webster Dictionary* (Springfield, MA: Merriam-Webster, 2005).

59. Macknik, Martinez-Conde, and Blakeslee, *Sleights of Mind*.

60. J. Otero-Millan et al., "Stronger Misdirection in Curved than in Straight Motion," *Frontiers in Human Neuroscience* 5 (2011): 133.

61. J. Cui et al., "Social Misdirection Fails to Enhance a Magic Illusion," *Frontiers in Human Neuroscience* 5 (2011): 103.

## CONCLUSION

1. J. R. R. Tolkien, *The Fellowship of the Ring: Being the First Part of "The Lord of the Rings,"* 2nd ed. (Boston: Houghton Mifflin, 1986).

# INDEX